D1017407

SEE YA LATER CALCULATOR

SIMPLE MATH TRICKS YOU CAN DO IN YOUR HEAD

BY THE EDITORS OF PORTABLE PRESS

PORTABLE PRESS
SAN DIEGO, CALIFORNIA

INTRODUCTION

uldn't be a four-letter word! It's useful, astounding, (gulp!) fun, yet four in ten Americans still claim " the subject. We figured that if we compiled t tricks, games, and do-it-yourself projects, we solve people's irrational fear of math. From tips e in everyday life to "special occasion" algebra etry, *See Ya Later Calculator* has everything master this radical subject. With these easy ven our most numerically challenged readers to do math in their head. Plus, the Pen + Paper n has bonus problems that are too hard to do less you're a math genius (in which case, you our next book!).

moguls, whose input was integral:

rroll	JoAnn Padgett
Deja	Megan Todd
rdjevic	Anna Nguyen
nssen	Alex Firer
n Dyl	Jay Newman
mas	Tracy Vonder Brink
one	Adam Bolivar
field	Tanya Fijalkowski
ger	Lilian Nordland
man	Meighen Severance

SEE YA LATER CALCULATOR

SIMPLE MATH TRICKS YOU CAN DO IN YOUR HEAD

Portable Press
An imprint of Printers Row Publishing Group
10350 Barnes Canyon Road, Suite 100, San Diego, CA 92121
www.portablepress.com · e-mail: mail@portablepress.com

Printers Row Publishing Group is a division of Readerlink Distribution Services, LLC. Portable Press is a registered trademark of Readerlink Distribution Services, LLC.

All correspondence concerning the content of this book should be addressed to Portable Press, Editorial Department, at the above address.

Publisher: Peter Norton
Publishing/Editorial Team: Vicki Jaeger, Tanya Fijalkowski, Lauren Taniguchi, Aaron Guzman
Editorial Team: JoAnn Padgett, Melinda Allman, J. Carroll, Dan Mansfield
Production Team: Jonathan Lopes, Rusty von Dyl

Cover design by Tom Deja
Illustrations by Ana Djordjevic and Rusty von Dyl
Interior design by Trina Janssen and Lidija Tomas

Library of Congress Cataloging-in-Publication Data
Names: Portable Press (San Diego, Calif.)
Title: See ya later calculator.
Other titles: See you later, calculator.
Description: San Diego, CA : Portable Press, 2017.
Identifiers: LCCN 2016005541 | ISBN 9781626867581 (hardcover)
Subjects: LCSH: Mathematics. | Mathematical recreations.
Classification: LCC QA95 .S425 2017 | DDC 510--dc23
LC record available at https://lccn.loc.gov/2016005541

Printed in United States of America
First Printing: June 2017
21 20 19 18 17 1 2 3 4 5

CONT

Math sh
and ever
they "hat
the nifties
could help
you can us
and geom
you need t
shortcuts,
will be able
Math sectio
mentally—u
should write

Our math

J. Ca
Tom
Ana Djo
Trina Ja
Rusty vo
Lidija To
Brian Bo
Dan Mans
Vicki Jae
Melinda Al

1. GENERAL MATH

EASILY TELL IF A NUMBER IS DIVISIBLE BY 3

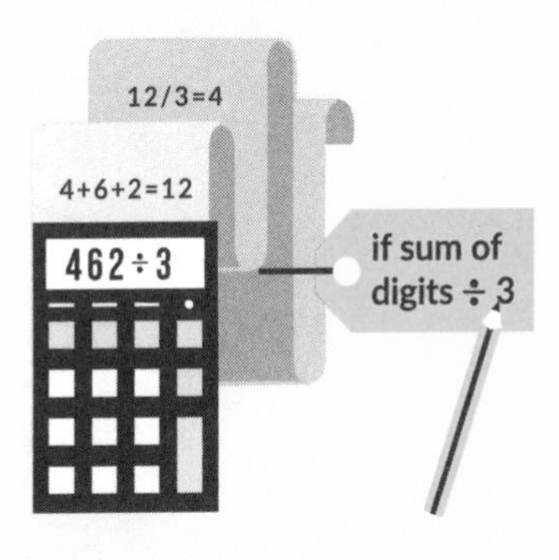

ADD BY ROUNDING

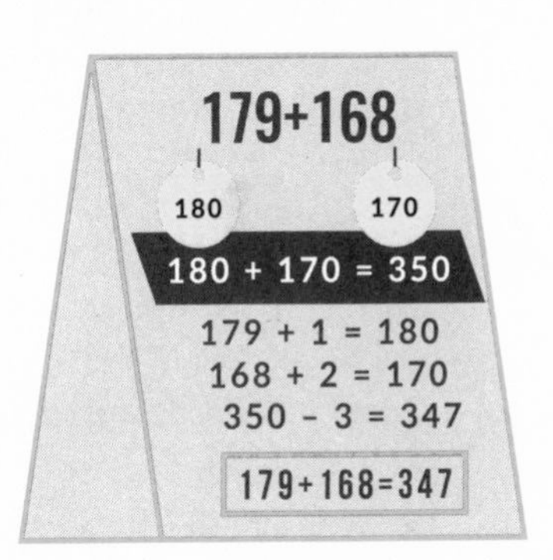

MULTIPLY BY SUBTRACTING

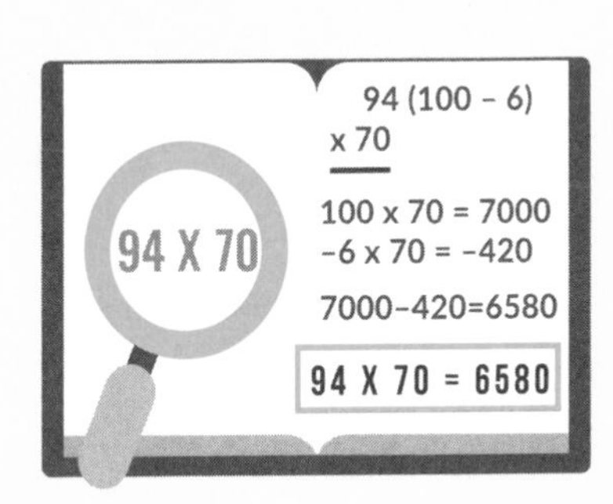

SQUARE ANY NUMBER

closest round # before

14 → 10

add amount subtracted to original #

14 + 4 = 18

18 x 10 = 180

$4^2 = 16$

180 + 16 = 196

ADD BY ALTERATION

Unlike your dry cleaners, we do alterations...but with math. (This is a math book, after all, not a dry cleaners.)

1 This trick helps make numbers more manageable and easily addable, so you can add them in your head. You just have to adjust the other number to account for any alterations you do on the first.

$$38 + 154$$

2 Get one of the numbers up to a nice, round "decade" number, like 20, 30, etc. Do this with simple addition.

$$38 + 2 = 40$$

3 Because 2 was added to one of the numbers, the same amount must be subtracted from the other number.

$$154 - 2 = 152$$

4 Add the two altered numbers.

$$40 + 152 = 192$$
$$38 + 154 = 192$$

ADD BY COLUMN

Here's a technique for adding a bunch of numbers without carrying over from one long column of digits to another—because not only is that annoying but it's hard to do without a piece of paper in front of you.

$$72 + 45 + 53 + 21$$

1 Add the digits in the "ones" column together.

$$\begin{array}{r} 7\mathbf{2} \\ 4\mathbf{5} \\ 5\mathbf{3} \\ +\ 2\mathbf{1} \\ \hline 11 \end{array}$$

2 Add the "tens" column digits together. Place it below the sum of the other column, but move it over one placement space.

```
   72
   45
   53
+  21
-----
   11
  18
```

3 Insert a "0" at the end of the second number, and add those two numbers together.

```
   72
   45
   53
+  21
-----
   11
+ 180
-----
  191
```

"Arithmetic is being able to count up to twenty without taking off your shoes."

—Mickey Mouse

PRE-DETERMINE YOUR GROCERY STORE TOTAL

You're at the grocery store. You're on a budget, or the store takes cash only. That means you're watching almost every cent—you'll need to estimate as you go.

Say you've come to the store to purchase exactly seven things: bread, eggs, milk, carrots, peanut butter, cereal, and a banana. You've got $20. Because you don't have a calculator or paper with you (you've got your list memorized!), you'll have to remember the prices along the way.

The trick is to "guesstimate" (that's guess + estimate) the prices by rounding up or down. If you do it to the nearest 50 cents (meaning if something is $0.20, round it down to 0; if something is $2.45, round up to $2.50), you'll be very close to the actual figure at the end—and within your budget. Plus the numbers will be easy to remember as you work your way down your list.

Item	Actual cost	Guesstimated cost
Bread	$0.79	$1.00
Eggs	$2.29	$2.50
Milk	$3.89	$4.00
Carrots	$1.99	$2.00
Peanut butter	$3.09	$3.00
Cereal	$3.39	$3.50
Banana	$0.29	$0.50
	$15.73	$16.50

Nice going! Your guesstimate was just 77 cents from the total, and well within your spending limit. That means you can buy everything on your list...as well as pay the sales tax.

SIMPLIFIED SUBTRACTION

If two-digit subtraction is hard, it's because things aren't simplified far enough. Here's a way to make it way simpler, way earlier.

The key is to break the numbers into manageable chunks.

$$\begin{array}{r} 474 \\ -\,148 \\ \hline \end{array}$$

1 Take the bottom number and round it down to a "decade" number and its remainder.

$$148 = 140 + 8$$

2 Now just subtract the first number from the top number, then the second number from that.

$$\begin{array}{r} 474 \\ -\,140 \\ \hline 334 \end{array}$$

$$\begin{array}{r} 334 \\ -\quad 8 \\ \hline 326 \end{array}$$

$$474 - 148 = 326$$

EASILY ADD UP THREE-DIGIT NUMBERS

You probably learned to do this kind of problem by starting at the far right, and working your way left, "carrying" extra digits as you go.

$$\begin{array}{r} 324 \\ +\,456 \\ \hline \end{array}$$

There's an easier way: Work from *left to right*. As you do, you break down the numbers into more mentally addable chunks.

1 Leave the top number alone, and then break down the bottom number into its hundreds column (400), its tens column (50), and ones column (6). Start by adding that top number to the bottom number's "hundreds" value.

$$\begin{array}{r} 324 \\ +\,400 \\ \hline \end{array}$$

2 Take that sum and now add the next value over, which is the "tens" column.

$$\begin{array}{r} 724 \\ +\ 50 \\ \hline 774 \end{array}$$

3 Only one column left: the far right or "ones" column. Take the sum from the previous step, and add what's in that spot.

Then proceed with that 6.

$$\begin{array}{r} 774 \\ +\ 6 \\ \hline 780 \end{array}$$

A 9-year-old boy named Milton Sirotta invented the word "googol" to stand for 10^{100}, or 1 followed by 100 zeros. (It can also be called ten duotrigintillion.) The name of the company Google came from an unintentional misspelling of "googol."

SUBTRACT A BIG NUMBER FROM 1,000

Here's a way to "eliminate" the carrying process when subtracting.

$$\begin{array}{r} 1{,}000 \\ -\ \ 245 \\ \hline \end{array}$$

1 Start by subtracting the first two digits of the second number from 9.

$$9 - \mathbf{2} = 7$$
$$9 - \mathbf{4} = 5$$

2 Subtract the last and final digit not from 9, but from 10.

$$10 - \mathbf{5} = 5$$

3 That's it. You're done. No carrying the 1 or any of that nonsense. Just line up the digits.

$$\begin{array}{r} 1{,}000 \\ -\ \ 245 \\ \hline 755 \end{array}$$

SUBTRACT BY ADDING

This neat trick works when subtracting two numbers on either side of 100.

$$\begin{array}{r} 187 \\ \underline{-\ 93} \end{array}$$

1 Determine how much each number is less than and more than 100. In this case, 187 is 87 more than 100, and 93 is 7 less.

2 Add 87 and 7, and you get 94, which is the answer to 187 - 93.

THE POWER OF 9

Here's a nifty mathematical secret: every number that's larger than 9 can be added up (or down) to 9. You can use it to win a bar bet, perhaps.

Every number greater than 9 becomes 9. How? Add up the digits of the number, and subtract that sum from the original number. The answer is always 9 or a number that has digits that add up to 9.

23

$$2 + 3 = 5$$

$$23 - 5 = 18$$

$$1 + 8 = 9$$

It works on big numbers, too—you just keep adding the digits and breaking them down until you finally get to 9.

525,603

$$5 + 2 + 5 + 6 + 0 + 3 = 21$$

$$525{,}603 - 21 = 525{,}582$$

$$5 + 2 + 5 + 5 + 8 + 2 = 27$$

$$2 + 7 = 9$$

SUBTRACT LARGE NUMBERS

Subtracting is easy. You just turn it into addition.

$$\begin{array}{r} 1{,}341 \\ \underline{-\ 488} \end{array}$$

Where do you even start with this? Right, you pretend that the "1" on the far right is an 11, and then you put a three down below the line, and then you carry the…blah. You can't do that in your head, and it's a pain to do on paper as well.

1 Round up the bottom number, the one that's being subtracted. (In other words, add.) Complete the easy subtraction problem you've created.

$$\begin{array}{r} 1{,}341 \\ \underline{-\ 500} \\ 841 \end{array}$$

2 But in adding extra to the bottom number, you've actually subtracted too much. To correct that, add to the solution however much you rounded up by.

$$488 + 12 = 500$$

$$841 + 12 = 853$$

$$\begin{array}{r} 1{,}341 \\ -\ 488 \\ \hline 853 \end{array}$$

Here's another example.

$$\begin{array}{r} 974 \\ -\ 586 \\ \hline \end{array}$$

1 Round up the bottom number. Solve.

$$\begin{array}{r} 974 \\ -\ 600 \\ \hline 374 \end{array}$$

2 Add the number you rounded by.

$$586 + 14 = 600$$

$$374 + 14 = 388$$

$$\begin{array}{r} 974 \\ -\ 586 \\ \hline 388 \end{array}$$

COMPLEMENTS OF 100

"My, what a strong round number you are!" Sorry, that's a *compliment* of 100. This is something different.

One way to do more complex math in your head is to have a bunch of shortcuts memorized. One such tool: complements of 100. This will help you find how far away a number is from 100 (which comes in handy for tricks that require rounding to the nearest 100). For example:

* The complement of 56 is 44 (because 56 + 44 = 100).
* The complement of 87 is 13 (because 87 + 13 = 100).
* The complement of 49 is 51 (because 49 + 51 = 100).
* It's also true in reverse — 56 is the complement of 44, and so on.

While these could be something you memorize (at least memorize the complements of round numbers like 10 and 20) to speed up your mental math, you can also quickly figure them out.

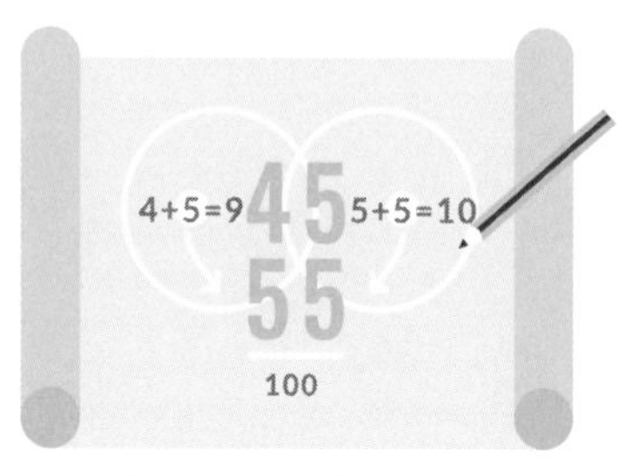

To determine a complement of 100, look at the left-hand number and find what number can be added to it to result in a sum of 9. Then look at the right-hand number and find what number can be added to make a sum of 10. (The numbers in their entirety add up to 100 because the "1" from that "10" gets carried, turning the 9 into a 10, and the result into 100.)

$$45$$

What number added to 4 makes 9? It's 5.

$$\begin{array}{r} 45 \\ +\mathbf{5} \\ \hline 9 \end{array}$$

And what number added to 5 makes 10? It's 5 again.

$$\begin{array}{r} 45 \\ +\mathbf{55} \\ \hline 10 \end{array}$$

$$\begin{array}{r} 45 \\ +55 \\ \hline 100 \end{array}$$

That means 45 and 55 are complements.

ADD BY ROUNDING

Line up the two long numbers, add the digits in each column, remember which digits you're carrying—who has the time? Here's a better way to add up big numbers. The secret is "round" here somewhere...

$$\begin{array}{r} 179 \\ +\,168 \\ \hline \end{array}$$

It wouldn't take long to add these, but it's a little tricky to keep the columns in place in your head and carry two digits. The trick is to round up or round down the numbers, and then account for the rounding later on.

1 Round up the first number.

179 becomes 180

2 Now do the same with the other.

168 becomes 170

3 You're left with two numbers that are easy to add.

$$\begin{array}{r} 180 \\ +\,170 \\ \hline 350 \end{array}$$

But this wasn't the original equation. To get the correct answer, you have to undo the adding you did when you rounded the numbers up. And what's the opposite of addition? Subtraction.

$$179 + 1 = 180$$
$$168 + 2 = 170$$

Since you added a total of 3 in the beginning, you have to subtract 3 now.

$$350 - 3 = 347$$
$$179 + 168 = 347$$

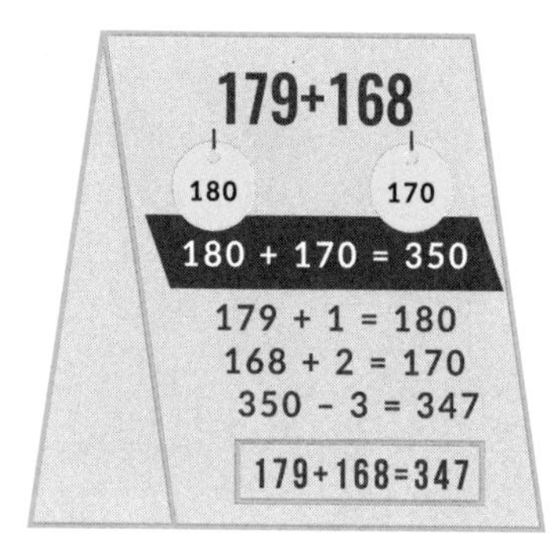

Here's an example of a problem that involves both rounding up and rounding down:

$$\begin{array}{r} 313 \\ +\,159 \\ \hline \end{array}$$

1 Round up or round down the numbers as necessary.

313 becomes 310
159 becomes 160

2 Add the simplified numbers together.

$$310 + 160 = 470$$

3 Account for the rounding.

$$313 - 3 = 310$$
$$159 + 1 = 160$$
$$-3 + 1 = -2$$

-2 rounding $= +2$ to solution

$$313 + 159 = 472$$

"Nature is written in mathematical language."
–Galileo Galilei

MATH MAGIC TRICK #1

Ready to become a powerful mathemagician, like Harry Potter, but with multiplication instead of actual magic? We've sprinkled these amazing math tricks throughout the book.

1 Ask someone to think of a number. Tell them to keep it to themselves.

2 Tell them to double the number.

3 Now have them add 12.

4 Divide by 2.

5 Subtract the initial number.

6 The number they've arrived upon, that they are thinking of now without telling you...is 6. The answer is always 6!

MULTIPLY NUMBERS THAT BOTH END IN ZERO

All those zeros really add up fast. Wait, that doesn't make any sense, does it?

3,000 x 400

1 Ignore all those zeros for a second and just focus on the other numbers. Multiply those together.

3 x 4 = 12

2 At the end of that number, tack on all of the zeroes you ignored in step one. Make sure to get all of them.

12 000 00
1,200,000

Let's do another one, for practice.

400,000 x 9,000,000,000

4 x 9 = 36

36 00000 000000000

3,600,000,000,000,000

COOL DIVISION TRICKS

Go forth and multiply. Or, you know, divide!

* If a number ends in 0, it's divisible by 10. To divide a 0-ending number by 10, just scratch out the 0.

$$70 / 10$$
$$7$$

* If you add up the digits of a number and you get 9, the number is divisible by 9.

$$243 / 9$$
$$2 + 4 + 3 = 9$$

* This works for numbers longer than three digits. If the last three digits is a number that's divisible by 8, the full number is also divisible by 8.

$$12{,}448 / 8$$
$$448 / 8 = 56$$
$$12{,}448 / 8 = 1{,}556$$

* If the last three digits of a number are 000, it's divisible by 8.

$$2{,}000 / 8 = 250$$

★ If a number is even and its digits add up to a number divisible by 3, then it's also divisible by 6.

$$642 / 6$$

(Yep, 642 is even.)

$$6 + 4 + 2 = 12$$

12 is divisible by 3 and 6.

$$642 / 6 = 107$$

★ If it ends in a 0 or a 5, a number is divisible by 5.

$$110 / 5 = 22$$
$$180 / 5 = 36$$
$$1{,}005 / 5 = 201$$
$$1{,}010 / 5 = 202$$

★ If its digits add up to a number that's divisible by 3, then it's divisible by 3.

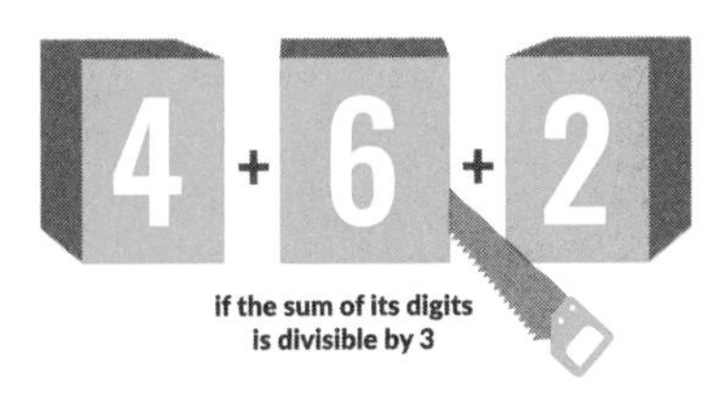
4 + 6 + 2
if the sum of its digits is divisible by 3

★ If it's an even number, it's divisible by 2. (That's what an even number is.)

SPLIT THE DIGITS

Sometimes everyone is just going to be happier if you just split up. Here's how to multiply two numbers by breaking them down.

$$35 \times 8$$

1 Take the larger, two-digit number and split it up into two factors—numbers that multiply against each other to get that number. Try to get one of the factors to be as small as possible, like a 2 or 3.

$$35 = 5 \times 7$$

2 Restate the problem, so you don't lose any of numbers.

$$8 \times 35 = 8 \times 5 \times 7$$

3 Now, take that new three-step multiplication problem (of smaller, more manageable numbers) and start working your way through

$$8 \times 5 = 40$$

$$40 \times 7 = 280$$

(Utilize another trick here: if 4 x 7 = 28, then 40 times 7 just gets a 0 at the end.)

$$35 \times 8 = 280$$

SIMPLE SIMPLIFY

Division problems are just fractions. This means you can use factoring to simplify them into smaller, easier-to-manage numbers.

387 / 81

A number whose digits add up to 9 are divisible by 9 (see page 18), and both of these numbers fit the criteria.

387 --- 3+8+7 = 18 --- 1+8 = 9
81 --- 8 + 1 = 9

Simplify the problem like you would simplify a fraction.

387 / 9 = 43
81 / 9 = 9
43 / 9

Depending on what your situation requires, you can now estimate the answer.

43 is almost 45
45 / 9 = 5
387 / 81 = about 5

If you put pen to paper, you get a more accurate solution.

45 / 9 (or 387 / 81) = 4.777777...

MULTIPLY A TWO-DIGIT NUMBER BY 11

Yes, you could do this kind of problem with a calculator, but the method below is actually faster. (Also, doing complex math in your head plays surprisingly well at parties.)

$$42 \times 11$$

1 Add up the individual digits in the other number in the problem, or the one that isn't 11. (In this case, that number is 42.)

$$\begin{array}{r} 4 \\ +2 \\ \hline 6 \end{array}$$

2 Now, take the sum and place it in between the digits of the original number. That's the solution.

4(6)2
462

MULTIPLY ANY NUMBER BY 5

Counting with your fingers is one way to do this, with each digit representing 5. But most of us have only 10 fingers, so the hands are useless beyond 10 x 5. Here's a quick tip for bigger numbers.

1,680 x 5

1 Take the number (the one that isn't the 5) and cut it in half.

$$\begin{array}{r} 1{,}680 \\ \div\ 2 \\ \hline 840 \end{array}$$

2 Place a 0 at the end

840 --- 0 --- 8,400

That's your answer.

1,680 x 5 = 8,400

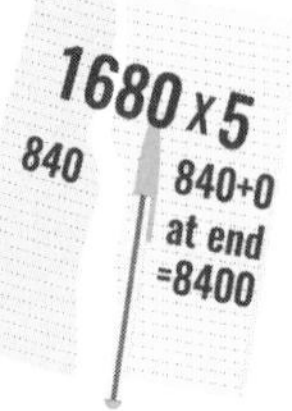

But what if the number you're multiplying is odd?

2,331 x 5

1 Subtract 1 from the number. That makes it an even number now.

2,331 - 1 = 2,330

2 Cut that number in half.

$$\begin{array}{r} 2{,}330 \\ \underline{\div 2} \\ 1{,}165 \end{array}$$

3 Add the zero at the end.

11,650

4 The 1 that was initially subtracted now has to be accounted for. But don't add 1 to the total—multiply that leftover 1 by 5 first.

1 x 5 = 5

5 Add the 5 to the total.

11,650 + 5 = 11,655

EASILY MULTIPLY BY 15

This is one of those problems that looks challenging at first glance...until you know how to break it down.

73 x 15

1 Take the number (the one that isn't the 15) and add a 0 to the end. (You're actually multiplying by 10 here, but don't think about that. It just needlessly complicates things.)

73 --- 730

2 Got the new number? Divide it in half.

$$\begin{array}{r} 730 \\ \underline{\div 2} \\ 365 \end{array}$$

3 Add those figures together.

$$\begin{array}{r} 730 \\ \underline{+\,365} \\ 1{,}095 \end{array}$$

$$\begin{array}{r} 73 \\ \underline{\times 15} \\ 1{,}095 \end{array}$$

MULTIPLY ANY NUMBER BY 9

The key is to exploit the fact that 9 is so very close to 10, an easy number by which to multiply. It's okay to exploit it, though. 9 is fine with it.

9 x 1,264,281

1. Add a zero to the number you're multiplying by 9. (Or if you prefer, multiply it by 10. It's the same thing.)

1,264,281 —> 12,642,810

2. Subtract the original number from it. (For really big numbers, you might need pencil and paper.)

$$\begin{array}{r} 12{,}642{,}810 \\ -\quad 1{,}264{,}281 \\ \hline 11{,}378{,}529 \end{array}$$

It works because you've just multiplied by 10, and then backtracked a little.

SPLITSVILLE

Like so many other things in this book, the key to making long division easier is to break down your numbers into smaller numbers that you can deal with in your head.

$$176 / 8$$

1. Split the number to be divided (the dividend) into numbers that can be evenly divided by the divisor.

$$176 = 96 + 80$$

2. Divide each number by the divisor.

$$96 / 8 = 12$$
$$80 / 8 = 10$$

3. And then you add those results.

$$12 + 10 = 22$$

That's your answer: 22.

Extra credit! Here's another way to look at the problem, if it makes more sense to you:

$$176 / 8 = (96 / 8) + (80 / 8)$$

DIVIDE AND CONQUER

Adding, subtracting, and multiplying big numbers in your head may seem easy compared to dividing big numbers in your head, but here are some methods to make long division more manageable.

1 **If possible, make a "guesstimate."** You don't need a calculator if you don't need an exact figure. For example, you have 166 rocks to place in three garden beds, and they don't have to be exactly even. Simply think of a number close to 166 that is a multiple of 3. The equation that should pop into your head is:

$$3 \times 50 = 150$$

You might also think of:

$$3 \times 60 = 180$$

So you can make a quick approximation that you'll need somewhere between 50 and 60 rocks for each bed.

2 **Break it down.** We began that process above by coming up with 3 x 50 = 150. If you want an exact number of rocks in each bed, then use a modified version of the trick we learned on the previous page. Subtract 150 from 166 to get 16. Now it's easy to divide that by 3, since

3 x 5 = 15. Add those two answers together—50 and 5—then you'll know pretty quickly that each garden bed should get 55 rocks. (You decide what to do with the one rock left over.)

3 **Go big.** The same process works for even larger numbers. If you have to divide 476 by 63, break both numbers into big, round numbers first by rounding each one up or down. That gets you 480 and 60. Take off the zeroes, and 48 / 6 = 8, which is your approximate answer. (To get the exact answer, use the trick on page 59.)

Farmer: "How many sheep do I have?"

Sheepdog: "You have 50."

Farmer: "That's odd. I thought I only had 47."

Sheepdog: "You do, but I rounded them up."

MULTIPLY BY SUBTRACTING

Here's another nifty way to make multiplication of two-digit numbers more approachable. This time you manipulate your numbers with a bit of simple subtraction.

$$\begin{array}{r} 94 \\ \underline{\times\ 70} \end{array}$$

1 Think of the top number not as itself, but as a certain distance from another, more easy to multiply number.

$$\begin{array}{rl} 94 & (100 - 6) \\ \underline{\times\ 70} & \end{array}$$

2 You've turned the number into a subtraction problem, or rather two numbers: one positive and one negative.

$$100 - 6$$
$$100 \text{ and } -6$$

3 Back to the problem, multiply each of those new figures by the other number in the original equation.

$$100 \times 70 = 7{,}000$$
$$-6 \times 70 = -420$$

4 Now take those two figures and solve the equation that resulted.

$$7{,}000 - 420 = 6{,}580$$
$$94 \times 70 = 6{,}580$$

Get some more practice with another example.

$$\begin{array}{r} 78 \\ \underline{\times\ 35} \end{array}$$

$$78 \quad (80 - 2)$$

$$80 \times 35 = 2{,}800$$
$$-2 \times 35 = -70$$

$$2{,}800 - 70 = 2{,}730$$
$$78 \times 35 = 2{,}730$$

SQUARE ANY NUMBER THAT ENDS IN 1

That's right, we're going to tell you how to quickly find out the solution for multiplying a number by itself. Any number at all...as long as it ends in 1. Hey, that's 10 percent of all numbers! (We just did another math trick right there—that one's free.)

$$41^2$$

1 Squaring is really just a multiplication problem, so restate it as such.

$$41 \times 41$$

2 Now, find the two whole numbers on either side of the number you're multiplying, and multiply those together. (You can use a calculator if you haven't read much else of this book yet. Several articles in this section show you how to multiply these numbers together in your head.)

$$40 \times 42 = 1{,}680$$

3 You've got to account for that 1 you rounded by. Square the 1 and add it to the total...which is easy, because 1^2 is 1.

$$1{,}680 + 1 = 1{,}681$$

$$41^2 = 1{,}681$$

Here's another bit of practice.

$$61^2$$

1 Restate as a multiplication problem.

$$61 \times 61$$

2 Take the whole numbers on each side of the number and multiply them.

$$62 \times 60 = 3{,}720$$

3 Account for the 1 by adding it to the solution.

$$3{,}720 + 1 = 3{,}721$$

$$61^2 = 3{,}721$$

SQUARE ANY NUMBER THAT ENDS IN 9

Is this the "Number 9 Dream" the Beatles were talking about on *The White Album*? Sure, why not. Here's how to find a number's square...if that number ends in 9.

$$19^2$$

1 Squaring is multiplying a number by itself, so restate it that way.

$$19 \times 19$$

2 Now, find the two whole numbers on either side of the number you're multiplying, and multiply those together.

$$18 \times 20 = 360$$

3 Add 1^2 to account for the 1 you left out when you changed the numbers in step 2. In other words, add 1.

$$360 + 1 = 361$$

$$19^2 = 361$$

SQUARE A NUMBER THAT ENDS IN 5

This one is *so* fast and easy, you'd swear it was magic, or at the very least, witchcraft.

$$25^2$$

1. Multiply the first digit of the number by its next sequential number (in other words, the next one).

$$\begin{array}{r} 2 \\ \times\ 3 \\ \hline 6 \end{array}$$

2. If a number ends in 5, its square will always end with "25." That's because 5^2 (or 5 x 5) = 25.

3. Take the result of step 1, and place it in front of the result of step 2.

$$6\ \ 25$$
$$25^2 = 625$$

That's all!

SQUARE ANY NUMBER

Squaring a number that ends in 1, 5, or 9 is super easy (see the previous pages). Use this trick for all the rest.

$$14^2$$

1 Find the closest round number that comes before the number.

$$14 \rightarrow 10$$

2 Take the amount subtracted to get to that round number, and add it to the original number.

$$14 + 4 = 18$$

3 Multiply that result by 10.

$$18 \times 10 = 180$$

4 Take that initial difference once more, and square it. Then add it to the result of step 3.

$$4^2 = 16$$
$$180 + 16 = 196$$
$$14^2 = 196$$

ESTIMATE HOW MUCH A STAR IS OVERPAID

When you hear that Adam Sandler earned $20 million for *Jack and Jill* (a movie so bad that it "won" every category at the 2012 Razzie Awards for terrible movies) or boxer Floyd Mayweather makes $30 million in a single fight, you can't help thinking celebrities are overpaid. But can you figure in your head *how much* they're overpaid? Sure!

As of 2016, the average Major League Soccer player makes an annual salary of $309,000. How much is that per week? It requires long division, but don't worry, it's not hard.

$$309{,}000 / 52$$

First, estimate the magnitude of the answer—whether it's hundreds, hundreds of thousands, millions, or whatever. Try a few guesses until you get an answer in the range of 309,000, but not over it. Let's try 10,000 first.

52 weeks x 10,000 = 520,000

That's too high. Try 1,000.

52 x 1,000 = 52,000

Close enough. Now to get a better estimate, drop the last several digits of each number you're dividing until you have a simple division problem that makes sense. Looking at just the first digit, 3 divided by 5 isn't easy. So look at 30 divided by 5 instead.

309,000 / 52,000

How many times does 5 go into 30?

It's about 6. Now multiply that by the estimated 1,000 you got earlier, and you get about $6,000 per week.

What about *really* big numbers? For instance, Sandra Bullock pulled in a whopping $75 million in 2016 for acting in the blockbuster *Gravity*. How much did she earn per day?

75,000,000 / 365

First, the magnitude. Try 10,000.

$$365 \text{ days} \times 10{,}000 = 3{,}650{,}000$$

Too low. What about 100,000?

$$365 \times 100{,}000 = 36{,}500{,}000$$

That's the right range. Look at the first two digits of each number. How many times does 36 go into 75?

$$75 / 36$$

It's about 2. Now multiply that by the 100,000 from earlier and you get $200,000 a day.

Extra credit: Compare that to your own pitiful salary, and try not to get depressed about your career choices.

DIVIDING BY FOUR

Let's say you and three friends just won a pie-eating contest. Congrats! You get to split the prize money of $128 four ways. But you won't need to pull out your phone and type the numbers in your calculator. Dividing a number by four is as easy as dividing it by two—you just have to do it twice!

128 / 4

1 First, divide 128 by 2. To make it even easier, use another trick to break it down: you know that 100 divided by 2 is 50, and that 28 divided by 2 is 14. Add those two answers to get 64.

2 Just divide 64 by 2 again, and you get 32 bucks...and your friend gets 32 bucks, and your other friend gets 32 bucks, and your other friend gets 32 bucks.

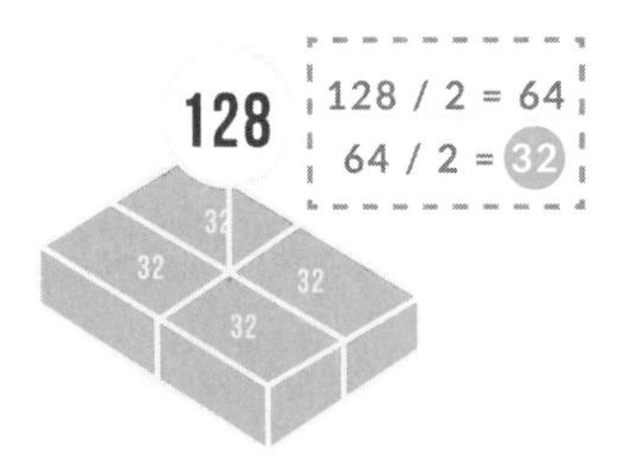

Now you can go out and buy...anything but more pies.

MULTIPLYING TEENS

This is about easily doing math problems involving numbers between 13 and 19. (There isn't some horrible place somewhere where teens are spontaneously cloning themselves. Whew!)

16 x 13

1 Add either number to the "ones" digit (the place farthest to the right) of the other number.

16 x 1(3)

$$\begin{array}{r} 16 \\ +\ 3 \\ \hline 19 \end{array}$$

2 Take that number and multiply it by 10.

$$\begin{array}{r} 19 \\ \times 10 \\ \hline 190 \end{array}$$

3 Now, multiply the "ones" place value of the original two numbers together.

$$\begin{array}{r} 6 \\ \times 3 \\ \hline 18 \end{array}$$

4 Add the two solutions you came up with together.

$$\begin{array}{r} 190 \\ +18 \\ \hline 208 \end{array}$$

Bonus: As is true with teen people, numbers that aren't quite teenagers yet still behave like teenagers. (By which, we mean this trick works with 11 and 12, too.) Take a look:

18 x 12

1 Add the first number to the ones place value of the other number.

18 x 1(2)

$$\begin{array}{r} 18 \\ +2 \\ \hline 20 \end{array}$$

2 Multiply that by 10.

$$\begin{array}{r} 20 \\ \times 10 \\ \hline 200 \end{array}$$

3 Multiply the ones values of the original problem

$$\begin{array}{r} 8 \\ \times 2 \\ \hline 16 \end{array}$$

4 Add the two solutions.

$$\begin{array}{r} 200 \\ +16 \\ \hline 216 \end{array}$$

It would take you about 12 days to count to one million. Counting to one billion would take you about 32 years.

MULTIPLY TWO NEARLY IDENTICAL NUMBERS

Here's a way to use an algebraic expression to multiply two large numbers together, quickly and easily.

Say you need to multiply two large numbers together, and they both so happen to be an equal distance apart from a whole, round number.

$$53 \times 47$$

In this case, that whole, round number is 50. They're both 3 units away from 50. (That's what "equal distance from a whole, round number" means.)

$$53 - 3 = 50$$

$$47 + 3 = 50$$

Now round them both to 50 and multiply, since that's an easier problem.

$$50 \times 50 = 2{,}500$$

To account for the amount you rounded, multiply that number by itself. In this case, you rounded by 3.

$$3 \times 3 = 9$$

Subtract the second answer from the first.

$$2{,}500 - 9 = 2{,}491$$
$$53 \times 47 = 2{,}491$$

Let's do it again with another example.

$$31 \times 29$$

Each number is 1 away from 30.

Round them both and multiply.

$$30 \times 30 = 900$$

And now multiply the amount you rounded by.

$$1 \times 1 = 1$$

Subtract.

$$900 - 1 = 899$$
$$31 \times 29 = 899$$

MULTIPLY TWO-DIGIT NUMBERS

This one seems pretty sneaky. But it works. And you'll save some paper.

$$21 \times 31$$

1 Multiply the first digit of both numbers together.

$$2 \times 3 = 6$$

2 Multiply the second digit of the numbers together.

$$1 \times 1 = 1$$

3 We're going to "cross-multiply." Take the first digit of the first number, and multiply it by the second digit of the second number. Then do the same thing with the other numbers.

$$2 \times 1 = 2$$
$$1 \times 3 = 3$$

4 Add those solutions together.

$$2 + 3 = 5$$

5 The number discovered in step 1 becomes the left-most digit of the problem's solution. The number from step 2 becomes the right-most digit of the problem's solution. In the middle, between them? The sum of the cross-multiplying from steps 3 and 4.

6 5 1
21 x 31 = 651

However, if that middle number is more than one digit, you've got a problem. The solution: carry the extra.

43 x 21

1 Multiply the first digit of both numbers together, and then the second digit of both numbers.

4 x 2 = 8
3 x 1 = 3

2 Cross-multiply all the digits, and add them up.

4 x 1 = 4
3 x 2 = 6
4 + 6 = 10

3 Keep the right-most digit of the middle number, and carry the overage to the number to the left.

8 10 3
903
43 x 21 = 903

AN AVERAGE METHOD

There are lots of times you might want to know an average—how much each person in a big group owes for a shared check, or how much you spend each month on restaurants, for example. Here's an easy way to estimate it.

To find an average, you add up every instance of an event, and then divide by the number of instances.

Amounts spent on restaurants over the last nine months:

$$140 + 174 + 210 + 98 + 146 + 155 + 156 + 201 + 204 = 1{,}484$$

$$1{,}484 / 9$$

If that's too hard, why not estimate the average? Since 9 is close to 10, you could average that number instead. Taking it one step farther, you could round up the other figure, too.

$$1{,}500 / 10 = 150$$

$$\$150$$

You spend roughly $150 per month. (That's not too far off from the exact answer, $164.89.)

VEDIC MATHEMATICS

This is a neat division trick supposedly invented by ancient Hindu clerics, but it wasn't "discovered" until 1965 when a modern Hindu cleric named Bharati Krishna Tirthaji introduced the concept in his book *Vedic Mathematics*.

681 / 5

It works like this: Let's say you want to divide 681 (the dividend) by 5 (the divisor). All you need to do is divide each of the numbers in the dividend by 5—and keep track of some numbers in your head.

1 First, divide 6 by 5, and you get 1 with a remainder of 1. Place that remainder before the next number, which is 8, and you get 18. 18 divided by 5 = 3, with a remainder. (Now you know the first two numbers of your answer will be 1 and 3.)

6 / 5 = 1, remainder 1

18 / 5 = 3, remainder 3

18 / 5 = 3, remainder 3

31 / 5 = 6, remainder 1

2 Now you have a remainder of 3. Place that before the final number, which is 1, to get 31. And 31 divided by 5 is 6, with a remainder of 1. So, place that 6 after the first two numbers of the answer (1 and 3).

681 / 5 = 136 with a remainder of 1

"When people say 'I hate math' what you're really saying is, 'I hate the way mathematics was taught to me.' Imagine an art class in which they teach you only how to paint a fence or wall, but never show you the paintings of the great masters. Then...you would say, 'I hate art.' What you would really be saying is 'I hate painting the fence.' And so it is with math. When people say 'I hate math' what they are really saying is 'I hate painting the fence.'"

–UC Berkeley math professor Edward Frenkel

MULTIPLY LEFT TO RIGHT

Multiplication is almost universally taught as a right-to-left process, starting with the ones column, proceeding to the tens, and so on. But mentally and verbally, math is expressed from left to right. Here's how to multiply the "right way."

$$\begin{array}{r} 51{,}231 \\ \times\ 3 \\ \hline \end{array}$$

1 Take the singular, bottom number and multiply it against each digit in the bigger, top number. Start on the left.

$$3 \times 5 = 15$$
$$3 \times 1 = 3$$
$$3 \times 2 = 6$$
$$3 \times 3 = 9$$
$$3 \times 1 = 3$$

2 Now, line up the digits. Problem solved.

153,693

That was a nice, clean example, with single-digit numbers in the middle. So what if those individual multiplication figures are more than 10? You carry the number, and add it to the preceding digit on the left. Here's how.

$$\begin{array}{r} 61{,}413 \\ \times\ 4 \\ \hline \end{array}$$

1 Once more, multiply each digit in the top number by the bottom number individually, working from left to right.

4 x 6 = 24
4 x 1 = 4
4 x 4 = 16
4 x 1 = 4
4 x 3 = 12

24,416,412

2 Uh oh, that's far too much. Estimation tells us the solution should be somewhere in the 240,000 range, not the 24,000,000 ballpark. What happened? You've got too many digits, and you have to carry some ones—on the two-digit solutions in bold.

3 Keep the digit in the ones spot and add the carryover to the preceding number.

24,416,412

245,652

HOW TO SQUARE THREE-DIGIT NUMBERS

Here's a unique way to multiply a large, three-digit number by itself—no complicated, multistep multiplication or memorization of big figures necessary!

$$204^2$$

1 Take the first and third digits of the three-digit number.

2 4

2 Now, find the square of each of those two digits.

2^2	4^2
4	16

3 Take each of the pulled-out digits, multiply them by each other, and multiply that by 2.

$$2 \times 4 \times 2$$
$$= 16$$

4 Place that figure between the initial squares you did up in step two.

4 16 16

41,616

In that example, the final "in between number" was two digits, so the problem was solved nice and cleanly. But if that number is three digits, you're going to have to carry a number. Here's how to do it.

825^2

1 Separate the number, but with the first digit by itself and the two other digits grouped together, like so:

8   25

2 Find the square of those two numbers.

8^2	25^2
64	625

3 You can use only the last two digits of that three-digit result here. Hold onto that 6.

64	(6)25
64	25

4 Multiply the separated-out numbers by each other, and by 2.

$$8 \times 25 \times 2 = 400$$

5 Insert the last two digits of that number between the squares from step 2.

6
64 00 25

6 As for the other digit, carry it over.

4 6
64 00 25

7 Add down your columns.

680,625

CHAPTER WRAP-UP

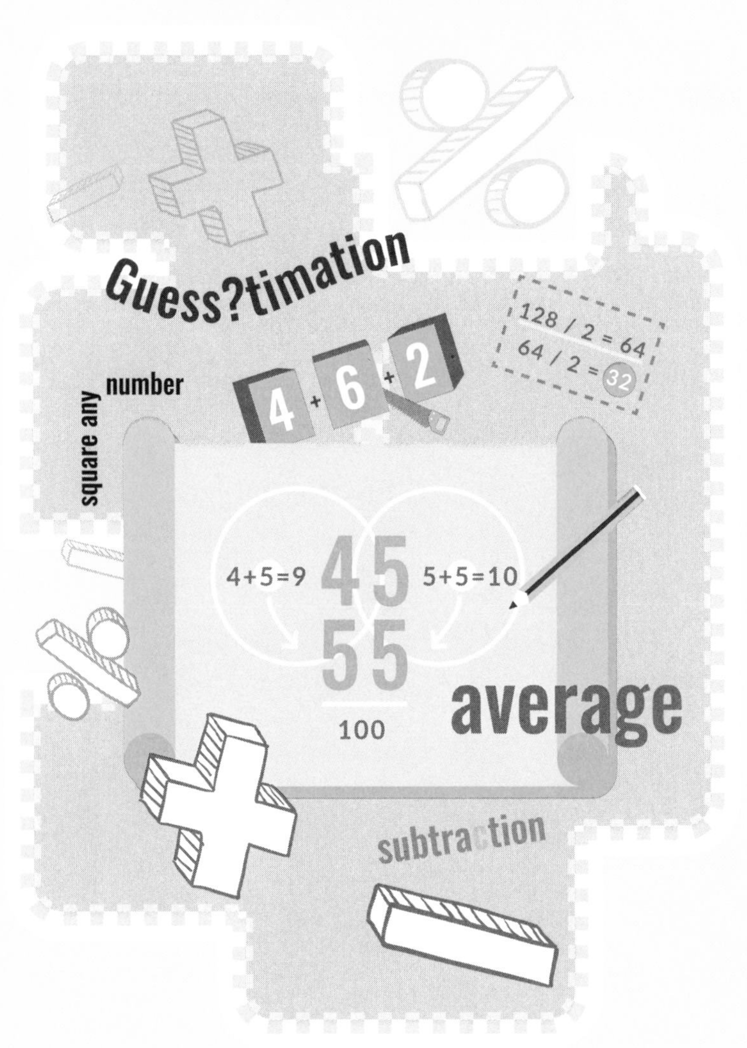

2. FUN WITH FRACTIONS

ADD MIXED NUMBERS
$1\frac{1}{2} + 2\frac{1}{3}$
$1\frac{1}{2}$
1 x 2 = 2
2 + 1 = 3
$2\frac{1}{3}$
2 x 3 = 6
6 + 1 = 7
$\frac{3}{2} + \frac{7}{3} = \frac{23}{6}$
3 x 6 = 18,
remainder of 5
$3\frac{5}{6}$

CONVERT REPEATING DECIMALS INTO FRACTIONS

0.7777777777...

1 — X = 0.7777777777...

2 — 10X = 7.777777777...

3 — 10x = 7.77...
\- x = 0.77...
9x = 7

4 — $\frac{9x}{9} = \frac{7}{9}$

$X = \frac{7}{9}$

KNOW YOUR ODDS

FIGURE TIPS AND TAX

FOR A 15% TIP

Move decimal point over one space to the left, then add half of the new number to itself.

1 — \$12.62 → \$1.26

2 — \$1.26+ 0.63= \$1.89

CALCULATE A TIP

For standard lunch, dinner, and tightwad rates.

For a 10 percent tip, or "cheapskate level":

Find the total amount of the check and move the decimal point to the left.

$12.62—> $1.26

For a 15 percent tip, or standard lunch tipping:

Move the decimal point over one to the left then add half of the new number to itself.

$12.62—> $1.26+ 0.63= 1.89

For a 20 percent tip, for dinner service:

Just get the 10 percent and multiply it by 2.

$12.62—> $1.26 x 2= $2.52

PERCENTAGE TRICKS

Here are some tricks to keep in mind when you're working with percentages. (The most common percentage you'll deal with: tipping. See the previous article for that.)

Per cent means "per 100."

Thus "6 percent" means "6 per hundred." Broken down more, that means that 1 percent of a number is 1/100 of a number, or 0.01. With that in mind, here's a way to look at percentages as more familiar math problems:

To find 5 percent, divide the number by 20. (5 percent equals 1/20).

5 percent of 120 = 6

To find 10 percent, divide the number by 10. (10 percent equals 1/10).

10 percent of 540 = 54

To find 20 percent, divide the number by 5 (20 percent equals 1/5).

20 percent of 445 = 89

To find 25 percent, divide the number by 4 (25 percent equals 1/4).

$$25 \text{ percent of } 288 = 72$$

To find $33\frac{1}{3}$ percent, divide the number by 3 ($33\frac{1}{3}$ percent equals $\frac{1}{3}$).

$$33\frac{1}{3} \text{ percent of } 594 = 198$$

To find $66\frac{2}{3}$ percent, divide by 3, and then multiply by 2 ($66\frac{2}{3}$ percent equals $\frac{2}{3}$).

$$66\frac{2}{3} \text{ percent of } 90 = 60$$

You can also flip these around if you're more comfortable working with fractions. Just remember the equivalents, and you're all set.

PERCENT FLIPPING

Or "flipping the percent." Call it what you want. It works either way, just like percent flipping (or flipping the percent, if you prefer).

The percentage (x) of a number (y) is the same as the *y* percentage of an *x* number. It's a bizarre vagary of math, and an example of the perfect, symmetrical beauty of numbers.

For example:

97% of 3 = 2.91

3% of 97 = 2.91

12% of 41 = 4.92

41% of 12 = 4.92

Here's one way you can use this to your advantage. If you need to find a percentage of a number, try flipping the percent to see if you can come up with an easier equation.

Finding 28% of 50 seems difficult, so switch it to 50% of 28. Now you can easily come up with 14.

HOW MUCH IS THAT SALES TAX?

Here are two ways to determine how much sales tax adds to your purchase.

Sales tax is given as a percentage. One of the easiest percentages to figure out in your head is 10%—just slide the decimal point one place to the left. For example, 10% of 175.00 is 17.50. Many sales taxes are around 5%, so for a fast estimate of sales tax, take 10% of the price and then cut the answer in half. If you're buying a pair of shoes for $48.00, 10% of the cost is $4.80. Half of that is $2.40...so a 5% sales tax on $48.00 is $2.40.

If you'd like to be precise (and you know the exact sales tax percentage):

1. Convert the sales tax percentage to a decimal by moving the decimal point two places to the left: 3.4% becomes .034; 6% becomes .06, and so on.

2. Multiply the decimal by the purchase price to find the sales tax.

3. Add the sales tax to the retail price to determine the total.

Let's say you're buying your $48 pair of shoes in California, where the average sales tax is 8.48%.

1. Move the decimal point two spaces to the left, and 8.48% becomes .0848.
2. Multiply .0848 by $48 (okay, you might need to write out that part). Voilà, your sales tax is $4.07.
3. Add the $4.07 sales tax to the $48 sticker price, and you'll be paying $52.07.

It's worth knowing how to calculate sales tax because the pricier the item, the more the sales tax adds up. If you buy a $101,500 Tesla Roadster in California, sales tax costs you an extra $8,607.20...or you could move to a state that has no sales tax, like Oregon.

QUICKLY ADD FRACTIONS

Remember how in school you had to make sure the bottom number was the same before you could add fractions? This is a different way to do it.

$$\frac{1}{7} + \frac{4}{5}$$

1 The numbers have different bottom numbers, or denominators. Don't mess with doing math to give them the same one (finding a "common denominator"). Instead, multiply the denominators you've got. That will be the solution fraction's bottom number.

$$7 \times 5 = 35$$

2 Now, cross-multiply the top numbers (numerators) of each fraction with the other number's denominator.

$$1 \times 5 = 5$$
$$7 \times 4 = 28$$

3 Add those sums to get the solution's numerator...

$$5 + 28 = 33$$

$$\frac{33}{35}$$

SUBTRACT FRACTIONS THAT HAVE A 1 ON TOP

Here's how to subtract one fraction from another...if both have a numerator of 1.

$$\frac{1}{3} - \frac{1}{7}$$

As is the case with regular, non-fraction numbers, to subtract, you just do the opposite of what you do with addition. As you do in adding fractions (see the previous article), ignore the numerators and use only the denominators. The difference between two fractions with numerators of 1 is equal to:

$$\frac{\text{The difference of the denominators}}{\text{The product of the denominators}}$$

1. Subtract the denominators.

$$7 - 3 = 4$$

2. Multiply the denominators (the same numbers used in the subtraction).

$$7 \times 3 = 21$$

$$\frac{1}{3} - \frac{1}{7} = \frac{4}{21}$$

3 Simplify the fraction if necessary (which doesn't need to happen this time).

Let's give this one another shot.

$$\frac{1}{2} - \frac{1}{14}$$

1 Subtract the denominators.

$$14 - 2 = 12$$

2 Multiply the denominators.

$$14 \times 2 = 28$$

$$\frac{1}{2} - \frac{1}{14} = \frac{12}{28}$$

3 Simplify if necessary.

$$\frac{12}{28}$$

$$\frac{3}{7}$$

$$\frac{1}{2} - \frac{1}{14} = \frac{3}{7}$$

SUBTRACT FRACTIONS

With work, family, and community obligations, who has time to find common denominators?

$$\frac{3}{5} - \frac{1}{3}$$

1 Multiply the denominators to get the solution's denominator.

$$5 \times 3 = 15$$

2 Now, multiply the first (larger) number's numerator by the other (smaller) number's denominator.

$$3 \times 3 = 9$$

3 Multiply the second (smaller) number's numerator by the first (larger) number's denominator.

$$1 \times 5 = 5$$

4 Subtract the smaller solution (step 3) from the larger solution (step 2). That's the final numerator.

$$9 - 5 = 4$$

$$\frac{4}{15}$$

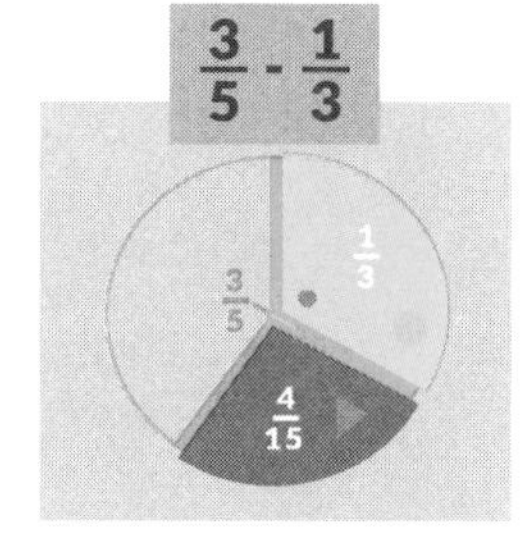

Let's try another example:

$$\frac{3}{4} - \frac{2}{3}$$

$$4 \times 3 = 12$$

$$3 \times 3 = 9$$

$$4 \times 2 = 8$$

$$9 - 8 = 1$$

$$\frac{1}{12}$$

MATH MAGIC TRICK #2

This one requires your volunteer to do all the math. Make sure he or she has a pen and a piece of scratch paper—not everybody has read this book, you know.

1 Have your volunteer write down a three-digit number, but here's the catch: the digit on the far left (the hundreds column) must be larger than the digit on the far right (the ones column). Tell the person to keep the number secret.

2 Have your volunteer reverse the number.

3 Subtract the reverse from the original number. (If that yields a two-digit number, have your volunteer put a 0 at the front.)

4 Now, take that result and add it to the reverse of itself. The answer is 1,089...and it will always be 1,089. No matter what.

Here's the math.

1 Choose a number where the first digit is larger than the final digit.

762

2 Reverse the number.

267

3 Subtract the reverse from the original.

$$\begin{array}{r} 762 \\ -\ 267 \\ \hline 495 \end{array}$$

4 Take the result from step number 3 and add it to the reverse of itself.

$$\begin{array}{r} 495 \\ +\ 594 \\ \hline 1{,}089 \end{array}$$

That's so weird, we've got to try it again!

$$\begin{array}{r} 473 \\ -\ 374 \\ \hline 99 \end{array}$$

$$\begin{array}{r} 099 \\ +\ 990 \\ \hline 1{,}089 \end{array}$$

ADD MIXED NUMBERS

A little bit whole number, and a little bit fraction, mixed numbers just can't be tamed. Or can they?

$$1\tfrac{1}{2} + 2\tfrac{1}{3}$$

1 You could turn them into decimals and add them that way. But that's hard to do in your head if you don't have the decimal equivalents of fractions memorized. (Which you don't. Because you have a life.) So, turn these mixed numbers into improper fractions. They're "improper" because the larger number is on top, not the bottom. Yeah, they get pretty crazy.

Multiply the whole number by the fraction part's denominator. Then, take that number and add the fraction's numerator. That sum is the improper fraction's numerator. The bottom number of the new, improper fraction is the denominator carried over from the fraction segment of the mixed number.

$$1\tfrac{1}{2}$$

$$1 \times 2 = 2$$

$$2 + 1 = 3$$

$$2\frac{1}{3}$$

$$2 \times 3 = 6$$

$$6 + 1 = 7$$

$$\frac{3}{2} + \frac{7}{3}$$

2 Add the two fractions, even though they're improper, as you would any two fractions. (For this, see page 74.)

$$\frac{3}{2} + \frac{7}{3} = \frac{23}{6}$$

3 Convert the improper fraction back into a mixed number. You do this by dividing the smaller, bottom number into the top number to find the whole number and then expressing the remainder as a fraction.

$$\frac{23}{6}$$

$3 \times 6 = 18$, remainder of 5

$$3\frac{5}{6}$$

MULTIPLY WHOLE NUMBERS BY FRACTIONS

Whole numbers. Fractions. They come from different worlds. Here's how to join them together in mixed number bliss.

$$15 \times \frac{2}{3}$$

1 Take the whole number, and then multiply by the numerator of the fraction.

$$\begin{array}{r} 15 \\ \times 2 \\ \hline 30 \end{array}$$

2 Take that number and divide by the denominator of the fraction.

$$\begin{array}{r} 30 \\ \div 3 \\ \hline 10 \end{array}$$

The answer: 10.

Now you know how to quickly find portions of whole numbers. This is a really good skill out in the real world.

Say you, your spouse, and your child are out to dinner with a friend, and you want to split the bill so that your family pays its share—3/4—and the friend pays the other 1/4. This method is how you'd figure out how to split the bill.

The problem: $\$56 \times \frac{3}{4}$

$$\begin{array}{r} 56 \\ \times 3 \\ \hline 168 \end{array}$$

$$\begin{array}{r} 168 \\ \div 4 \\ \hline 42 \end{array}$$

The answer: You pay $42. The friend pays the difference, which is $14.

But hang on, here's another interesting math anomaly. You can multiply a whole number by a fraction by starting at the bottom of the fraction instead of at the top. You just divide first and then multiply, instead of the other way around. Let's use the dinner example again.

$$\begin{array}{r} 56 \\ \div 4 \\ \hline 14 \end{array}$$

$$\begin{array}{r} 14 \\ \times 3 \\ \hline 42 \end{array}$$

CONVERT DECIMALS THAT REPEAT FOREVER

How you tease us, repeating decimals. The same number, or pattern of numbers, going on and on. Why can't you be clean and tidy, like fractions? (Pro tip: Grab a pencil for this one; it's probably too hard to do in your head.)

$$0.7777777777\ldots$$

1 Sorry, but your high school algebra teacher wasn't lying when she said you'd actually use algebra at some point. Here is that point. Convert this to an algebraic equation, and solve for *x*, in which x is the fraction we're after.

$$x = 0.7777777777\ldots$$

2 Multiply both sides by 10. That moves the decimal place by one, but important, position on the repeating decimal.

$$10x = 7.777777777\ldots$$

3 Here's where it gets weird. Subtract the algebraic expression from step 1 from the algebraic expression in step 2.

$$\begin{array}{r} 10x = 7.777777777\ldots \\ \underline{-\quad x} = \underline{0.7777777777\ldots} \\ 9x = 7 \end{array}$$

4 Divide the number off the x side to isolate the x (as one does in algebra), and make sure to do the same to the other side (as one does in algebra).

$$\frac{9x}{9} = \frac{7}{9}$$

$$x = 7/9$$

A similar method can be used if the decimal repeats a pattern, or the same sequence of numbers over and over and over and over...Again, solve for x.

$$x = 0.242424242424\ldots$$

1 Once again, solve for x.

$$x = 0.242424242424\ldots$$

2 Multiply both sides by as many powers of 10 that are necessary to move the decimal point to where the pattern starts to repeat. In this example, the pattern is 24 and starts repeating after two place values. That means multiply each side of the equation by 100.

$$100x = 24.2424242424\ldots$$

3 Subtract the algebraic expressions.

$$\begin{array}{r} 100x = 24.2424242424\ldots \\ \underline{-\quad x = 00.2424242424\ldots} \\ 99x = 24 \end{array}$$

4 Isolate the x, and simplify the fraction, if necessary.

$$x = 24/99$$
$$8/33$$

3 out of 2 people have trouble with fractions.

MATH MAGIC TRICK #3

Be it trickery, or be it math? (It be math.)

1 Think of a number. *Any* number. It doesn't matter what it is. Even though this was written months, even years before you're reading it, we'll guess what number you end up with.

2 Add 5 to your number. (Don't tell us what it is yet!)

3 Got it? Double the sum. (No. You can't tell us yet.)

4 Now subtract 4. (Keep it to yourself, man!)

5 Cut the number in half. (We'll tell you when it's time for the reveal.)

6 Do you remember the original, still secret number you started out with? Subtract it from the figure you came up with at the end of step 5.

7 The number you wound up with is 3. Told you we'd get it. But how did we know it would be 3? *Because the answer is always 3.*

WHAT ARE THE ODDS?

The terms "odds" and "probability" can be confusing, especially because they're often used incorrectly. Both words describe the chance that a specific event will occur, but there is a big difference:

* Probability expresses chance as the number of desired outcomes out of the number of total possible outcomes. It is expressed as either a fraction or a percentage.

* Odds measures the chances for and the chances against an event ever occurring. It is expressed as a ratio.

1 For example, if you flip a coin, the chances are that you will have one of two possible outcomes: heads and tails. So to find the probability of flipping heads, you divide the number of desirable outcomes (1) by the number of possible outcomes (2). So the probability of flipping heads is 1/2, or 50%.

2 Once you figure out probability, you can figure out the "odds in favor" or the "odds against" the desirable event occurring. When flipping a coin, you have one chance of getting heads, and one chance of getting tails. So the odds are expressed as a ratio: 1:1, which you would say as "1 to 1 odds."

PICK A CARD

Still with us? Now let's say you pick randomly from a deck of 52 cards. What is the probability that you'll get a heart? And what are the odds?

1 To find the probability of picking one of the 13 hearts, divide the total number of outcomes (52) by the number of desirable outcomes (13). That answer, expressed as a fraction, is 13/52, which can be simplified to 1/4. Probability is most commonly expressed as a percentage, so there is a 25% chance you'll pick a heart.

1:3
ODDS OF PICKING A HEART

2 To find the odds in favor of picking a heart, which in this case is your "desirable outcome," you subtract the number of desirable outcomes (13) from the number of total possible outcomes (52). That answer is 39, which represents the number of undesirable outcomes (picking a club, spade, or diamond). So to put that in "odds" terms, you use a ratio. In this case it's 13:39, which would be said as "13 to 39 odds in favor of picking a heart." (Notice how those two numbers add up to 52.) Of course, this being math, we would simplify that ratio to 1:3. Flip it around to get the odds against. "There are 3 to 1 odds against picking a heart."

3 What's the probability that you'll pick a 2? There are four 2's (desirable outcomes) out of 52 cards (total outcomes). So the probability is 4/52, or 1/13, which can be expressed as 0.08, or 8%.

4 To find the odds, subtract the number of desirable outcomes (4) from the number of total outcomes (52), which is 48, for a ratio of 4:48. Simplify that to 1:12. So the odds in favor of picking a 2 are 1:12. The odds against picking a 2 are 12:1.

"Math is the only place where truth and beauty mean the same thing."
—Danica McKellar

THE GAMBLER'S FALLACY

Is your "hot streak" really so hot?

You flip a coin and it comes up heads seven times in a row. Surely the eighth toss will come up tails because it's due. Right? Wrong! And that's the gambler's fallacy: mathematically, there's no such thing as "due."

Probability, the prediction of an event's likelihood, is calculated by the number of ways the event might occur divided by the total number of outcomes. In other words, a coin can land one way out of two possible outcomes. So, the probability of a toss landing heads up is, as is the probability of it coming up tails. It's the same 50-50 chance every time a coin is flipped, regardless of what has happened before. In dice, a single six-sided die can land 1, 2, 3, 4, 5, or 6, so the probability of any of those outcomes is 1/6.

Assuming the coin isn't rigged or the die isn't loaded, the probability of an outcome never changes. Instead of trying to predict the next throw based on a "run," you'd be better off remembering another gambler's saying: "The dice have no memory."

THE MONTY HALL PARADOX

Remember Monty Hall, the host of the TV game show *Let's Make a Deal*? Monty (and the show) may be gone, but his name lives on in a fascinating probability puzzle that was inspired by one of the "deals" on his show.

Imagine you're a contestant on *Let's Make a Deal*, standing in front of three giant doors labeled "1," "2," and "3." Behind one of the doors (you don't know which one, but Monty does) is a new car. Behind the other two doors are booby prizes: live goats. Monty Hall invites you to choose a door; you'll win whichever prize is behind it. You pick a door—say, Door #1. But before Monty tells you what you've won, he opens one of the doors you didn't pick, say Door #3, to reveal...a goat. Then he asks you, "Do you want to switch to Door #2?" Well, do you? Will switching from Door #1 to Door #2 improve your chances of winning the car? This puzzle, originally called the "Monty Hall Problem," was first proposed by a statistician named Steve Selvin in 1975.

If you think the odds are the same whether you stick with Door #1 or switch to Door #2, you're not alone. That's what most people would say, because that seems to make sense. After all, if there's one car and three doors, the odds of it being behind Door #1, Door #2, or Door #3 are exactly the same: 1-in-3. But that's the wrong way to look at the problem. According to Selvin, you have to think of it in terms of the one door you picked versus the two doors you didn't pick: The odds that the car is behind the door you picked are 1-in-3, and the odds that the car is behind one of the two you didn't pick are 2-in-3.

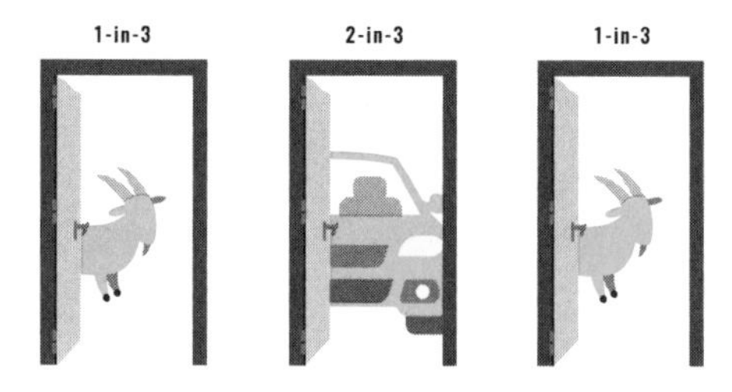

The odds don't change when one of the doors is opened because the prizes haven't moved. Sure, once Door #3 is opened to reveal a goat, the odds of the car being behind that door drop to zero. But there's still a 2-in-3 chance that the car is behind one of the two doors you didn't pick. That means there's now a 2-in-3 chance that the car is behind Door #2. Switching from Door #1 to Door #2 actually doubles your odds of winning the car—from 1-in-3, to 2-in-3. So switch doors!

A 20 PERCENT CHANCE OF RAIN

What does a weatherperson mean when she predicts a 20 percent chance of an afternoon shower? If you're like most people, you don't actually know.

Weather forecasts are often deemed a bust by the general public because they are misinterpreted. Often, people hear what they *want* to hear. If there's a 20 percent chance of rain, they may assume it won't rain and may be upset if it does. The consequences can range from merely bothersome (you plan a picnic based on what seemed like a good forecast, but it rains) to life threatening (being caught on a boat in a hurricane).

The average person doesn't have the same conception of weather and weather terms as the meteorologist does. Do you know what a 70 percent probability of rain means? In a national survey, only 10 percent of respondents knew. It tells us that on 7 days out of 10 with a similar weather pattern, a given place within the forecast area (your house or the airport, for example) will receive measurable rainfall. In simpler terms, a probability of precipitation forecast (POP) is just what it sounds like—an expression of the forecaster's confidence that there will be rain or snow. If there is good chance, the forecaster may say 80 percent; if dry weather is expected, the POP may be 20 percent. Officially, POP refers to the probability of at least .01 inches of precipitation at any given point in the forecast area over a specified time period. In the end, the value of an accurate forecast hinges on how well it is understood by the public.

Quick memory trick: You may not remember that there are 5,280 feet in a mile...and you don't have to. Just think of "5 tomato."

5--5
2--to
8--(m)ate
0--o

THE MARTINGALE

You might want to rethink that "foolproof" gambling system.

Nobody knows how the Martingale system got its name, but it has been in use since at least the 18th century. Renowned lover Casanova once wrote, "I went [to the casino of Venice], taking all the gold I could get, and by means of what in gambling is called the martingale I won three or four times a day." The Martingale is used in even-money games where the bettor wins or loses the same amount of money in each play. The system is simple: the player doubles the bet after every loss. That way, the gambler might win enough to make up for previous losses plus one. For example, if you start with a $1 bet and lose, on the next round you double your bet to $2.

$$1 \rightarrow 1 \times 2$$

Lose again, and your bet becomes $4.

$$2 \rightarrow 2 \times 2$$

If you win that hand, which doubles your money, you'll have $8, which covers the $4 you put into that hand as well as the previous $3 in losses, plus you'll have a profit of $1.

$$4 \rightarrow 4 \times 2$$

But does it work? Michael Shackleford, an actuary who analyzes casino games, created a computer program to find out. His simulation compared a Martingale bettor with a flat bettor wagering on the pass line in craps. The flat bettor bet $1 every time, win or lose, and the Martingale bettor doubled the bet after any loss. After running the simulation for 1,000,000 sessions of each type of bet, he found the Martingale bettor did no better or worse than the flat bettor!

$$1 \rightarrow 1 \rightarrow 1$$
$$=$$
$$1 \rightarrow 2 \rightarrow 4$$

Not only does the Martingale fail to give an advantage, it's also risky. Shackleford explains, "It is easier than you think to lose several bets in a row and run out of betting money after you've doubled it all away." So, throwing double the good money after bad is likely to leave you broke much faster.

CHAPTER WRAP-UP

3. ALL HAIL ALGEBRA

REMEMBER THE ORDER OF OPERATIONS

Parentheses
Exponents
Multiplication
Division
Addition
Subtraction

CALCULATE YOUR SALARY

How much is an hour of your time worth?

For every $1 per hour you earn, that's about $2,000 a year.

$41,000 / 2,000 = $20.50

FIND THE OUTSIDE TEMPERATURE USING A CRICKET

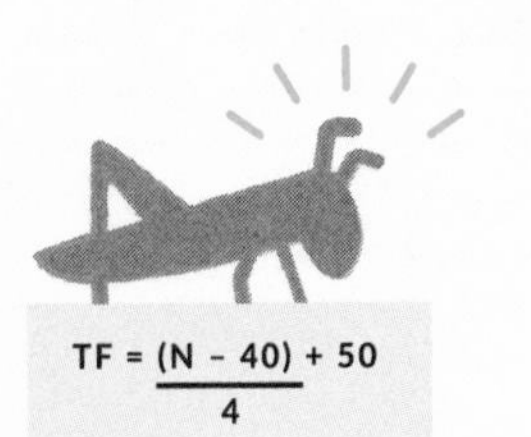

$$TF = \frac{(N - 40)}{4} + 50$$

TF is temperature (in Fahrenheit).
N is the number of chirps per minute.

ESTIMATE TIME AND DISTANCE TRAVELED

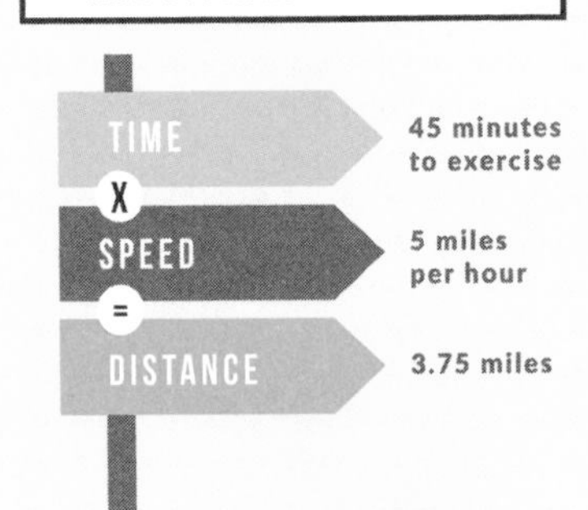

HOW TO REMEMBER THE ORDER OF OPERATIONS

Unlike the arts or philosophy, math is objective—there are definite right or wrong answers. And to get the right answer in algebra, you have to follow the "order of operations." Here's how.

A common mistake people make is they try to solve equations from left to right, since that's how we read. Instead, think of PEMDAS—or turn it into a mnemonic like this: "Please Excuse My Dear Aunt Sally." It stands for:

P: Parentheses

E: Exponents

M: Multiplication

D: Division

A: Addition

S: Subtraction

This is the order in which to do things in complicated equations that have bracketed expressions, numbers to the nth power, and so on. If one of the elements of PEMDAS isn't present in an equation, just skip to the next thing.

Here's an example, using (almost) every possible letter of PEMDAS.

$$1 + (3 \times 2)^3 / 4(30 + 6)$$

1. Do the *P*, by completing all tasks inside of parentheses.

$$(3 \times 2) = 6$$

$$(30 + 6) = 36$$

$$1 + 6^3 / 4(36)$$

2. Next, solve any exponential instructions.

$$6^3 = 216$$

$$1 + 216 / 4(36)$$

3. Now multiplication.

$$4 \times 36 = 144$$

4. Move on to division.

$$216 / 144 = 1.5$$

5. Now addition.

$$1 + 1.5 = 2.5$$

6. There's no subtraction here. The problem is solved.

TURN KILOMETERS INTO MILES

We figure there are two main situations when you might need to know how many miles are in a kilometer: If you're driving in metric Canada, or if you're a distance runner. Here's your cheat sheet.

One kilometer equals 0.62137 miles. Got it? Good.

Okay. Just kidding. Start with the kilometers you want to convert:

80 km

1 Cut it in half.

$$\begin{array}{r} 80 \\ \underline{\div 2} \\ 40 \end{array}$$

2 Divide that number by 4, and add it to the half.

$$\begin{array}{r} 40 \\ \underline{\div 4} \\ 10 \end{array}$$

$$\begin{array}{r} 10 \\ +\,40 \\ \hline 50 \end{array}$$

That's it! 80 km = 50 miles

(It's really about 49.7 miles, but close enough.)

Let's try another one. How many miles have you signed yourself up to run if you've entered a 10K race?

10 km

1 Cut it in half.

$$\begin{array}{r} 10 \\ \div\,2 \\ \hline 5 \end{array}$$

2 Take a quarter of that, and add it to the half of the starting figure.

$$\begin{array}{r} 5 \\ \div\,4 \\ \hline 1.25 \end{array}$$

$$\begin{array}{r} 1.25 \\ +\,5 \\ \hline 6.25 \end{array}$$

You're going to spend your Saturday morning running about 6.25 miles for some reason.

HOW TO FIGURE YOUR SALARY

Determining your yearly salary based on an hourly wage, or your hourly wage from your yearly earnings, is tough for anybody, unless your job is a math teacher.

1 Rule of thumb: For every $1 per hour you earn, that works out to $2,000 a year. (This assumes a 40-hour workweek, two weeks' vacation, and no overtime.) So, let's say you make $8 an hour.

$$8 \times 2{,}000 = 16{,}000$$

An $8-an-hour job nets $16,000 a year.

2 Curious about how much an hour of your time is really worth, and you're on a fixed salary? Just do the math backward: Divide your yearly pay by 2,000.

$$\$41{,}000 / 2{,}000 = 20.50$$

Someone who makes $41,000 a year is paid at an hourly rate of $20.50.

HOW LONG BEFORE YOUR INVESTMENT DOUBLES?

Want to get an idea of how good a potential investment might be...years before it pays off? Well, we can tell you. (We're talking savings and interest-bearing accounts—we can't predict the stock market. If we could, we wouldn't be writing math books.)

$$y = \frac{72}{r}$$

It's called the "Rule of 72." To find out how many years (y) it will take for compounding interest to double your initial investment, or the principal, divide the interest rate (r) into 72. You don't even need to put the principal, or initial investment, into the formula.

Principal: $5,000
Interest rate (r): 4 percent

$$y = \frac{72}{4}$$

$$y = 18$$

That means in 18 years, compounding interest will turn that $5,000 into $10,000.

The Rule of 72 only works if the interest rate is lower than 20 percent, which it's probably always going to be, unless you go to some mythical bank that pays you 21 percent interest on your savings account.

The method works backward, too, if you just want to double your money in a certain period of time. To do that, divide the period of time in which you'd like to have double the money into 72, and it will give you the interest rate you need to make it happen.

$$r = \frac{72}{y}$$

Let's say the time is 8 years.

$$\frac{72}{8} = 9 \text{ percent interest rate}$$

Q: What is Welsh mathematician Robert Recorde's claim to fame?
A: He invented the equals sign in 1557.

RUNNING HOT AND COLD

An American in Paris, you check the weather and see that it's 25 degrees. But that parka and sweater you're wearing prove unnecessary under the hot sun. What happened? You mistook Celsius for Fahrenheit.

$$\text{Celcius} \times \frac{9}{5} + 32 = \text{Fahrenheit}$$

1 The formula above is more exact, but here's a simpler way to change a Celsius temperature *C* into a more familiar Fahrenheit (*F*).

$$(C \times 2) + 30 = F$$

2 Plug the Celsius temperature into the formula.

$$(25 \times 2) + 30 = F$$
$$50 + 30 = 80$$
$$25 \text{ degrees C} = 80 \text{ degrees F}$$

1 Here's how to go the other way, should you be a European person in America:

$$(F - 30) / 2 = C$$

2 Plug in the numbers. For example:

$$(66 - 30) / 2 = C$$
$$36 / 2 = 18$$
$$66 \text{ degrees F} = 18 \text{ degrees C}$$

SIMPLE TIME MATH

Adding amounts of time together is just basic addition, right? Yeah, except that everything turns over to the next digit at 60 instead of 100. That makes it a bit more nerve-wracking to do time math. But not anymore!

Say you want to figure out how long it's going to take you to watch two movies. One of them is an hour and forty-five minutes long, and the other movie is two hours and ten minutes.

1 First express the lengths as units of time.

1:45
2:10

2 Add the hours together. And then the minutes together. But separately. And then just put them next to each other.

$$1 + 2 = 3$$
$$45 + 10 = 55$$

3:55

It will take you three hours and fifty-five minutes to watch both movies.

But whatever, that one was easy, because you had the hour and minutes to begin with, and the final minutes came in under 60. Let's do a time math problem that's a little more complicated.

Let's while away a Saturday watching the entirety of *The Godfather* trilogy. *The Godfather* has a running time of 177 minutes, *The Godfather Part II* runs 200 minutes, and *The Godfather Part III* is 162 minutes. How many hours will you be spending in front of the TV?

1 First, convert the minutes-only figures into hours-plus minutes. Do this by pulling 60 minutes—an hour—as many times as possible. Each 60 minutes equals an hour; the leftover minutes are then expressed as minutes (obviously).

The Godfather:

$$177 - 60 - 60 = 57$$

2 hours, 57 minutes

The Godfather Part II:

$$200 - 60 - 60 - 60 = 20$$

3 hours, 20 minutes

The Godfather Part III:

$$162 - 60 - 60 = 42$$

2 hours, 42 minutes

3 Now, express those times as hours and minutes with the colon between them.

$$\begin{array}{r} 2{:}57 \\ 3{:}20 \\ +\ \underline{2{:}42} \end{array}$$

Add up the hours, and then separately, add up the minutes.

$$2 + 3 + 2 = 7$$
$$57 + 20 + 42 = 119$$

$$7{:}119$$

4 Since minutes can only be expressed from 0 to 59, that 119 has to be converted to hours, which are then carried over to the hours spot. Do this by once again subtracting 60 from it as many times as possible.

$$119 - 60 = 59$$

5 Add the hours to the hours column, and leave the leftover minutes in the minutes column.

$$7{:}119$$

$$8{:}59$$

It will take you 8 hours and 59 minutes to watch *The Godfather* trilogy, or 9 hours with a 1-minute bathroom break. Have fun!

MATH MAGIC TRICK #4

The trick relies on the power of 9. (And a simple knowledge of the 9 times table. See? It's finally coming in handy!)

1 Ask someone to secretly write down any number that's at least four digits long. Let's say they choose 56,832.

2 Now ask that person to add up the digits, in this case, 5+6+8+3+2 = 24.

3 Have them subtract the answer from the first number (56,832 – 24 = 56,808).

4 Next, ask the person to cross out any one digit from the answer except a zero and then read you the new number. Let's say the 6 was crossed out, so the number now is 5,808.

5 Even though you haven't seen any of the numbers, you can tell the person the number that was just crossed out. They'll be amazed!

Here's how it works:

After the person has added up the digits and done the subtraction, the answer will always be divisible by 9. And if a number divides evenly by 9, when its digits are added up, they will also divide by 9. If you check our example—5+6+8+0+8 = 27—you'll see that it divides by 9.

So, after the person crosses out a digit and reads you the digits that are left, you add them up. In this case, 5+8+0+8 = 21. Then ask yourself, what's the next number that's divisible by 9? 27. What would you have to add to 21 to get 27? Six. If the numbers added up to 9, the next number divisible by 9 is 18, which is 9 away. Whatever number you get, that's the number that was crossed out!

"Here's some advice: At a job interview, tell them you're willing to give 110 percent. Unless the job is a statistician."

—Adam Gropman

GOING THE DISTANCE WITH FIBONACCI

This cool trick makes use of the Fibonacci sequence. Named after the 12th-century mathematician who popularized it, it's a series of numbers in which each entry is the sum of the two numbers that came directly before it. Here's a taste.

The Fibonacci sequence begins like this:

1, 1, 2, 3, 5, 8, 13, 21, 34, 55, 89, 144…

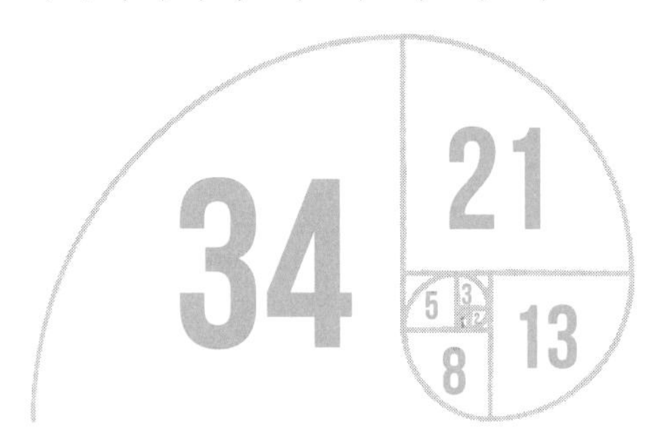

It has a lot of real-life applications…and here's one of them. You can use it to approximately convert miles to kilometers.

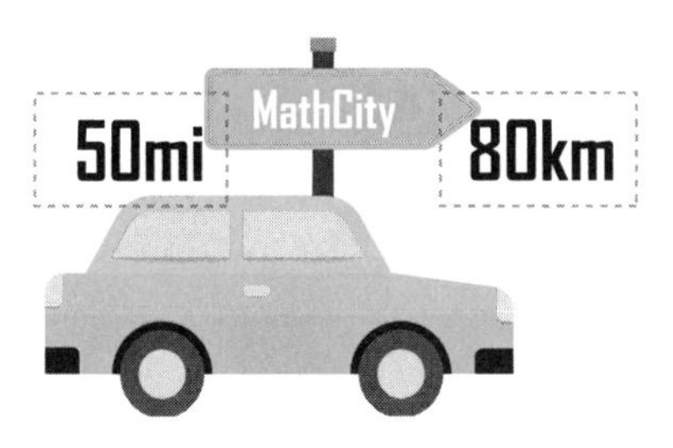

1 Take the number of miles you wish to convert, and add it to the previous number in the Fibonacci sequence.

5 miles

$$5 + 3 = 8$$

5 miles = 8 km (roughly)

Based on that ratio, you can convert bigger numbers, or ones that aren't a part of the Fibonacci sequence.

1 How many kilometers is 30 miles?

5 miles = 8 kilometers

30 miles = ?

2 Take the ratio between the 30 and 5, and apply the same to that 8 km.

$$30/\,5 = 6$$

$$6 \times 8 = 48$$

30 miles = 48 km (roughly)

HOW TO CALCULATE COMPOUND INTEREST

If you don't care so much about how long it takes for your investment to double (see page 106) and are just curious about how much interest you'll accrue over time, check out this simple formula.

To figure out how much interest you'll earn on an interest-bearing account, you've got to know a few things about the investment, which hopefully is information you have.

P: The principal, or amount of the initial investment

R: The interest rate, expressed as a decimal

N: How often interest is applied or compounded each year (generally monthly)

T: The investment period, or length of time you're measuring the interest-building (in years)

Those are used to create this formula, which will show you how much interest is earned over time.

$$a = p(1 + \frac{r}{n})^{nt}$$

Let's try it out. You've just deposited $8,000 in a savings account. It offers a 5 percent interest rate, compounded monthly. What will your account balance be in six years?

P = $8,000

R, expressed as a decimal = 0.05

N, or how often the interest compounds in a year = 12

T, in years = 6

$$8{,}000(1 + \frac{0.05}{12})^{12 \times 6}$$

Work through the problem, according to PEMDAS (see page 101).

$$8{,}000\ (1 + .004167)^{12 \times 6}$$

$$8{,}000\ (1.00416666)^{72}$$

$$8{,}000\ (1.349)$$

$$10{,}792$$

That $8,000 investment turns into $10,792 over time. Not bad!

KILOGRAMS VS. POUNDS

Here's another method to make those mystifying metrics more manageable.

$$p = 2k + 0.1(2k)$$

To convert from kilograms to pounds, doubling the number and adding a tenth of that number totally works. Doubling a number is easy, and getting a tenth (or 10 percent, or 1/10, or 0.1) is as easy as moving the decimal point over.

14 kilograms is equal to how many pounds?

1 First, double the number

$$14 + 14 = 28$$

2 Find a tenth of that figure.

$$28 \times 0.1 = 2.8$$

3 Add the two numbers together

$$28 + 2.8 = 30.8$$

(14 kilograms is actually 30.8647 pounds...but close enough!)

DISTANCE, SPEED & TIME

You've probably heard the expression "distance equals speed times time." But what does it really mean? And how can you use it in real life?

Distance = Speed x Time

This is an algebraic expression. That means that if you know two of the three elements, the third one can easily be determined with a simple bit of division.

In the most obvious real-world example: How long is it going to take you to drive somewhere? This formula can be used, provided that the speed, or rate, at which you travel is constant. It probably isn't—traffic is never consistent—but it will give you a pretty good idea.

Let's say you're headed over the river and through the woods to Grandmother's house for Thanksgiving. It's 160 miles away, and it's almost entirely freeway, which means you can travel at 65 miles per hour the whole time. You know distance and speed, so you'd solve the expression for time (t).

Pro tip: When setting up any equation, make sure all the elements are expressed in the same units. For example, if speed is in miles per hour, then distance must be in miles, and time must be in hours (not minutes). In this case, we've got miles and miles per hour, so we're all clear.

$$160 = 65t$$

Isolate the *t* to get your answer.

$$\frac{160}{65} = t$$

Simplify if needed.

$$t = 2\frac{30}{65}, \text{ or } 2\frac{6}{13}$$

It will take you about two and a half hours to get there.

Or say you've got 45 minutes to exercise, and you decide to run on a treadmill at a constant rate of 5 miles per hour. How far will you have run (distance, or *d*) in 45 minutes?

First, make your units consistent. The speed is in miles per hour, so the time needs to be converted to hours. To do that, divide the minutes by 60.

$$45 \text{ mins}/ 60 = 0.75 \text{ hours}$$

$$d = 5 \times 0.75$$

$$d = 3.75$$

You'll have run 3.75 miles.

THE CRICKET THERMOMETER

If you want to know the temperature and don't have a thermometer handy, all you need is this formula and a cricket...and a knowledge of insect calls.

$$TF = \frac{(N - 40)}{4} + 50$$

Dolbear's law allows us to use snowy tree crickets, known as "nature's thermometers," to tell the temperature to within 1°F. Found across much of the United States, these crickets are called *snowy* not because they live in snow but because they're so pale they appear to be almost white. After noticing that crickets chirp faster when it's warmer, American inventor Amos Dolbear first published the equation in 1897 (after he lost a telephone patent battle against Alexander Graham Bell).

Here's how the formula works:

TF is temperature (in Fahrenheit).

N is the number of chirps per minute.

Count the number of chirps in a minute, subtract 40, divide that number by 4, and add 50, and you've got the temperature.

$$TF = \frac{(N - 40)}{4} + 50$$

TF is temperature (in Fahrenheit).
N is the number of chirps per minute.

There's also a shortcut, which will give you an approximation. Just count how many times the cricket chirps in 15 seconds (which is the same as taking the number of chirps per minute and dividing by 4). Then add 40.

$$TF = \frac{N}{4} + 40$$

You can do this with other crickets, but each requires its own formula. For example, for the common field cricket, which gives a less accurate temperature but is easier to locate, add 38 to the number of chirps per 15 seconds (which, again, is the number of chirps per minute divided by 4).

$$TF = \frac{N}{4} + 38$$

HOW TO FIND YOUR BMI

To determine if you're at a healthy weight (or not), the National Institutes of Health recommends a look at your BMI, or "body mass index." It's a number based on your weight relative to height that provides a general idea about whether your weight is a good, healthy one. Here's how to figure it.

Even though the BMI is a standard calculation in the American medical community, it's derived from your weight in kilograms and height in meters, both metric figures. So first, let's convert.

1. Multiply your weight in pounds by 0.45. This will change the pounds to kilograms.

150 x 0.45 = 67.5 kilograms

2. Now, multiply how many inches tall you are by 0.025 to convert it to metrics. (To find out how many inches tall you are, multiply the number of feet tall you are by 12—because there are 12 inches in a foot—and then add the extra inches to that figure.)

5'8"

12 x 5 = 60

$$60 + 8 = 68 \text{ inches}$$

$$68 \times 0.025 = 1.7 \text{ meters}$$

3 Square your height in meters.

$$1.7^2$$

$$1.7 \times 1.7 = 2.89$$

4 Divide your weight in kilograms by your squared height. Round those extra numbers to one decimal place.

$$67.5 / 2.89 = 23.4$$

In this example, the body mass index for a 150-pound, 5'8" person is 23.4. Now compare this to the NIH's body mass index table:

BMI	Rating
Less than 18.5	Underweight
18.5– 24.9	Normal
25–30	Overweight
30+	Obese

Keep up those healthy habits, imaginary person whose stats were used in the example above: 23.4 is in the normal BMI range.

MATH MAGIC TRICK #5

Here's a calculator trick to try out on your friends.

1 Ask someone to think of a number between 1 and 100, and to keep it a secret.

2 Using a calculator, take your age and multiply by 2.

3 Add 5.

4 Multiply by 50.

5 Subtract 365.

6 Keeping that number on the calculator, hand the calculator to your victim, and tell him or her to add the secret number.

7 Then add 115.

Voilà! The first half of the resulting number is your age, and the other part of the number is your friend's secret number!

TRAINS OF THOUGHT

Let's say you have a secret lover and want to figure out when and where two trains traveling toward each other will meet so you can rendezvous. Here's how.

1 Train 1 is a bullet train that leaves San Francisco at 1:00 p.m. heading south at an average rate of 120 mph. Train 2 is a much slower freight train that leaves San Diego at the same time heading north, but it only averages 35 mph. Where and when will they meet? The formula is:

Distance = Rate x Time

2 What we don't know, however, is the time. So restructure the equation: Time = Distance / Rate.

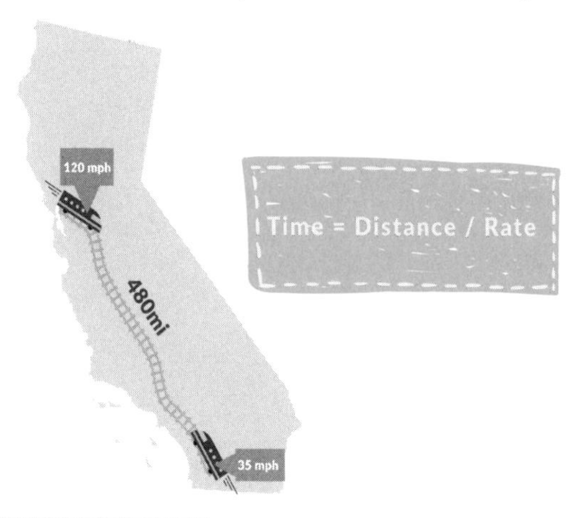

3 Let's plug in the numbers that we know. The two cities are roughly 480 miles apart, so that's the distance.

4 To find the rate, you need the relative rate, or how far both trains will travel in one hour. To get that, you simply add the two rates together: 120 + 35 = 155. (That's how fast the trains will appear to be going relative to each other when they meet.)

5 Now your equation looks like this: Time = 480 / 155. So using your new division skills, you know that the trains will meet in approximately 3.1 hours. To convert the decimal part of your answer to minutes, multiply it by 60. And that comes out to 6 minutes. They'll meet in 3 hours and 6 minutes, at 4:06 p.m.

6 To figure out where the trains will meet, go back to the original formula, Distance = Rate x Time. You can apply this formula to either train because they're both going to the same place. Let's use Train 1. First, make sure the rate and time are expressed in the same units. Since the rate is 120 miles per hour, the time should also be in hours. We know from step 5 that it's 3.1 hours.

Now plug the numbers in.

$$\text{Distance} = 120 \times 3.1 = 372$$

So the bullet train will have traveled 372 miles when it meets the freight train (in Los Angeles).

FREE FALLIN'

Galileo Galilei was the first to suggest that all objects fall at the same rate, no matter how much they weigh. And he was right!

According to legend, Galileo dropped two balls—one heavy and one light—from the top of the Leaning Tower of Pisa, and observed that they hit the ground at the same time. We do know that air resistance can interfere with the result. However, as astronaut David Scott demonstrated on the surface of the Moon, in a vacuum a feather will fall exactly as fast as a hammer. And how fast is that? On Earth, the rate is:

$$32.2 \text{ ft/s}^2 \text{ (or } 9.8 \text{ m/s}^2\text{)}$$

If you drop an object, after one second, it will be falling at 32.2 feet per second. After two seconds, it will be falling another 32.2 feet per second faster: 64.4 feet per second. After three seconds, it will be falling at 96.6 feet per second...and so on.

You can calculate the distance an object will fall in a given amount of time using a formula where *g* represents gravitational acceleration: 32.2 feet per second squared. The unit *t* is time in seconds, and *d* is the distance.

With a little algebraic jiggery-pokery, you can rearrange the equation to find out the time it will take an object to fall a given distance.

$$d = \frac{gt^2}{2}$$

$$2d = gt^2$$

$$\frac{2d}{g} = t^2$$

$$\sqrt{\frac{2d}{g}} = t$$

Say you're bungee jumping off a 150-foot-high bridge and your friend is waiting below to take a picture of you just as you reach the bottom. How will she time it just right so that she knows how many seconds after you jump off the bridge to take the picture?

Plug in 150 ft for *d* and 32.2 ft/s^2 for *g*.

$$t = \sqrt{\frac{2 \times 150}{32.2}}$$

You will get a time of about 3.05 seconds.

In this case, air resistance is a variable that's unaccounted for. The drag force air has on a falling body increases until it stops speeding up and reaches what's called "terminal velocity." On Earth, that means you can never fall faster than about 200 mph. Sorry, speed demons!

REAL-LIFE USES OF THE QUADRATIC EQUATION

The quadratic equation: it's not just for math geeks!

Simply put, a quadratic equation is an equation with at least one squared variable (like x^2). Quadratics have been used since ancient Babylonian times, although the standard form of the quadratic equation wasn't published until the 12th century. If you've forgotten, the standard quadratic equation is $ax^2 + bx + c = 0$ where *a*, *b*, and *c* are known values and *a* is not equal to 0. The solutions—there are almost always two—occur when the quadratic equation is equal to zero. This version of the formula is often used:

$$X = \frac{-b \pm \sqrt{b^2 - 4ac}}{2a}$$

The military uses quadratics to calculate where missiles will land, and engineers rely on quadratics in their designs. But what about the rest of us? It turns out that quadratics can be useful to you, too! For example, quadratics come in handy when calculating how to price an item to make a profit.

Let's say you want to sell cakes at a charity event, and you need to know how to price them to make money. Your start-up costs are $600 to buy kitchen equipment, and each cake has $2 worth of ingredients. So how much do you charge per cake? Your market research on similar events tells you every time the price of a cake goes up by $1, 20 fewer people buy one, and you know there will be 500 people at the event. If you call the price of each cake *p*, the formula for demand is:

$$\text{Number of sales} = 500 - 20p.$$

You also want to know how many dollars of sales to expect. The formula for that is:

$$\text{Sales in dollars} = \text{Number of sales} \times \text{price, or}$$

$$\text{Sales in dollars} = (500 - 20p) \times p\text{, or}$$

$$\text{Sales in dollars} = 500p - 20p^2$$

Of course, you also have your costs to consider. Remember you bought $600 in equipment, and each cake will cost $2 to make. Your formula for costs is:

Costs = Fixed costs + (2 x number of sales), which can be rewritten as:

$$\text{Costs} = 600 + (2 \times (500\text{-}20p))$$

$$\text{Costs} = 600 + 1{,}000 - 40p$$
$$\text{Costs} = 1{,}600 - 40p$$

Finally, the formula for profit is: Profit = Sales in Dollars – Costs. In this case, it's:

$$\text{Profit} = (500p - 20p^2) - (1600 - 40p)$$

$$\text{Profit} = -20p^2 + 460p - 1600$$

But why turn that into a quadratic? Solving the quadratic will tell you the two exact prices where the profit is zero. These are the points where you make no profit because the cakes are priced too low or because the cakes are priced too high. The best price is the one halfway between those two numbers. So, you're going to need to solve for the places where the profit is zero:

$$-20p^2 + 460p - 1600 = 0$$

Put that equation into the quadratic formula, and you get:

$$p = \frac{-460 \pm \sqrt{460^2 - (4)(-20)(-1600)}}{(2)(-20)}$$

The plus and minus signs mean that there are two equations to do here: one that uses addition and one that uses subtraction. That yields two possible answers:

p = 4.28 is the "too low" price, and p = 18.72 is "too high." Your optimal cake price is $11.50, exactly halfway between those two numbers.

Other real-life uses for quadratics include calculating speed—say a rower wants to know her speed going up and down a river. A football player could use quadratics to determine how long the ball will take to hit the ground after it's thrown. There are so many possibilities!

Having trouble remembering the quadratic formula? Sing it to the tune of "Pop Goes the Weasel":

X equals minus b
Plus or minus the square root
Of b-squared minus four ac
All over two a

CHAPTER WRAP-UP

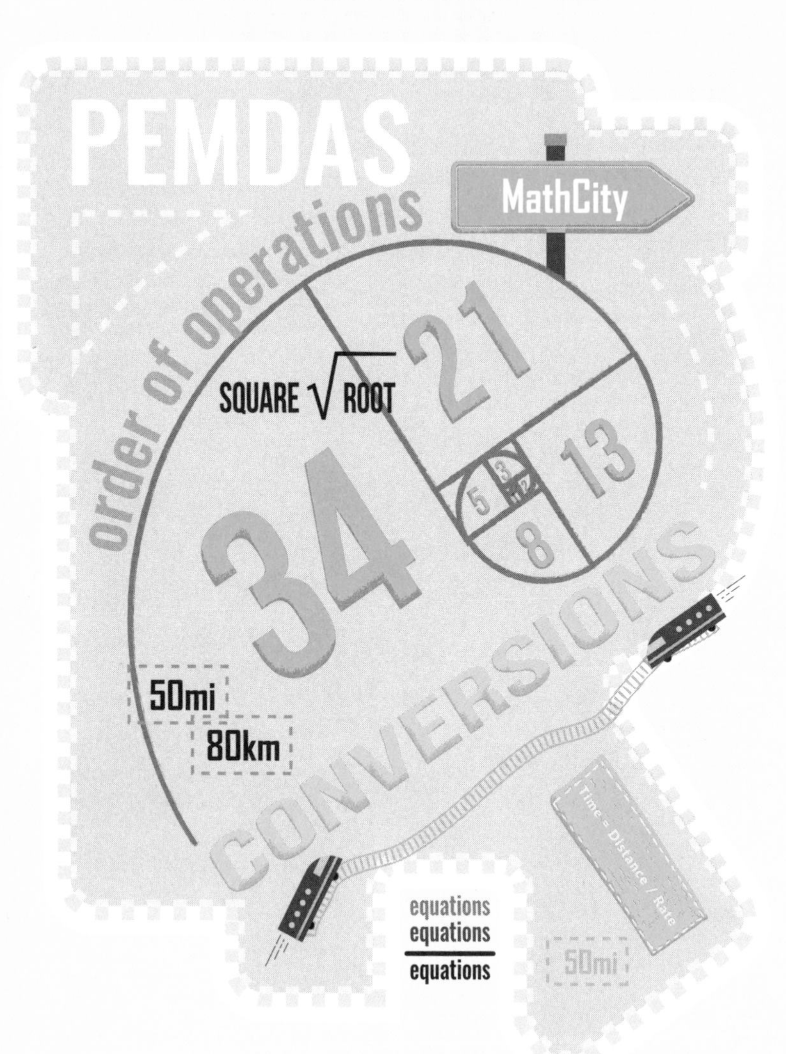

4. ENJOYABLE GEOMETRY

FIND THE AREA OF DIFFERENT SHAPES TO PAINT A ROOM

USE ANGLES IN REAL LIFE

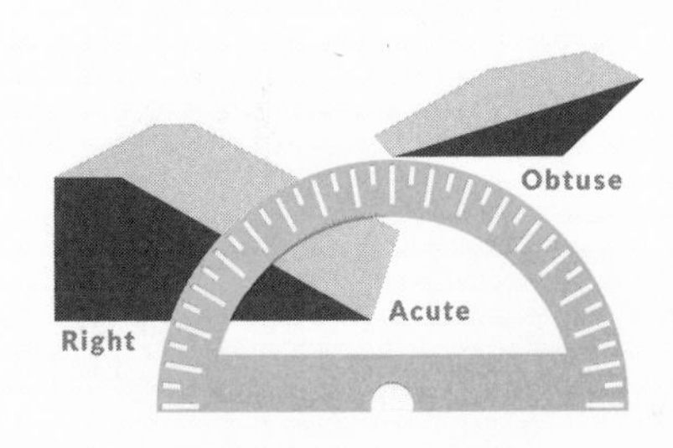

FIGURE THE CAPACITY OF A BARREL OF PUNCH

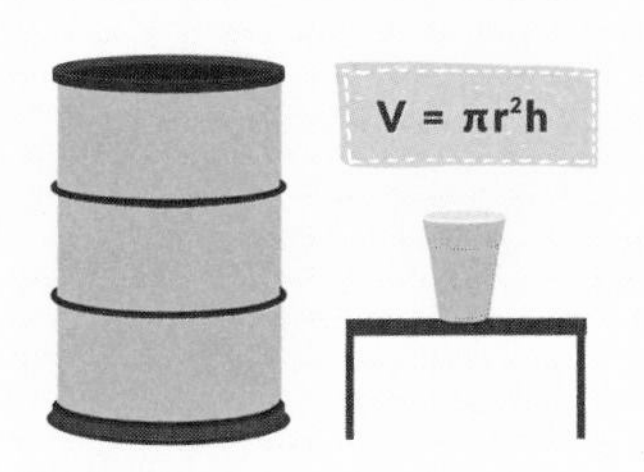

USE THE PYTHAGOREAN THEOREM FOR MEASUREMENT

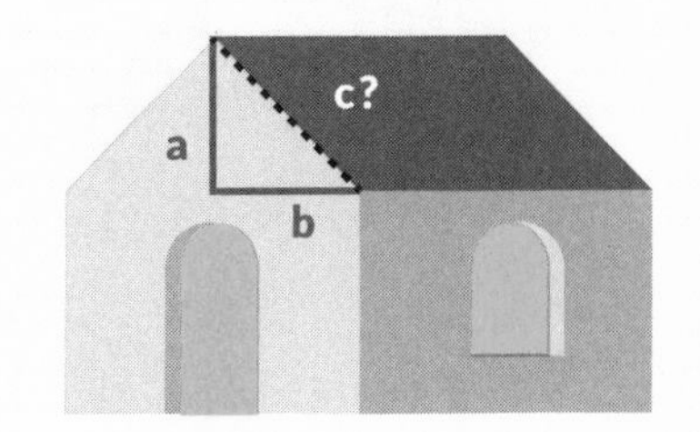

WRAP PRESENTS WITH GEOMETRY

Some people have a natural ability to wrap presents, and other people...don't. But fret not—you can use math to figure out how much paper you need to make square and rectangular boxes look pretty.

1 Measure the diagonal length across the largest side of the box. This will be *a*.

2 Find the height of the package (*b*).

3 Find half of *b*. This will be *c*.

4 Add a + b + c.

5 Cut out a square of wrapping paper with each side measuring equal to that number.

6 Place the present diagonally in the center of that paper (so that the corner of the package is pointed out toward the midpoint of a side).

7 Bring in all four corners together. They should meet at the center on the bottom of the package. The rest is up to you!

ANGLES IN REAL LIFE

There are many times in life when you'll need to calculate an angle. Here's how to estimate it. (For exact measurements, learn how to make your own protractor on page 203.)

There are three kinds of angles:

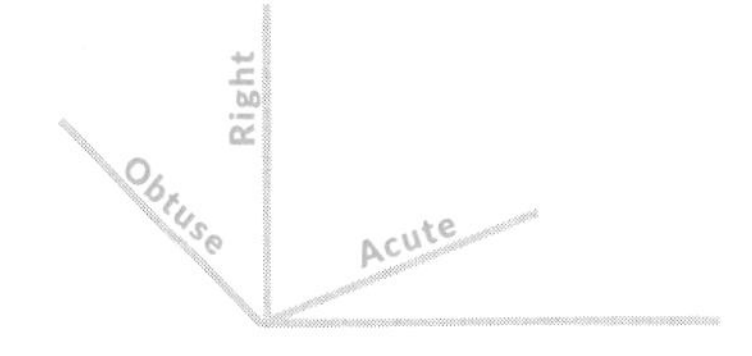

Acute: An angle less than 90 degrees.

Right: An angle of exactly 90 degrees (like the corner of a square).

Obtuse: An angle of more than 90 degrees but less than 180 degrees. (An angle of 180 degrees is a straight line.)

To determine the measurement of an unknown angle, use a right angle and a straight line as benchmarks. For example, a square's corner has a right angle of 90 degrees. Look at an angle and estimate the size by eyeballing it.

Does it look about halfway between the two sides of a right angle? Then it's a 45-degree angle (90 / 2 = 45). One-third of the way? 30 degrees. (90 / 3 = 30).

If the angle bends the other way (more than 90 degrees), you can determine its size the same way—just add 90. If it looks like it's going a third of the way past 90 degrees, it's a 120-degree angle. (90 + (90 / 3) = 120).

Here are situations in which it's important to know angles:

* To read a star chart, and determine the location of a star or constellation based on its angle relative to Earth.

* To know the downgrade of a hill, while riding a skateboard or driving, for example.

* To construct a house, particularly windows, doors, and roofs.

* For navigation, used by pilots and boat captains.

* To set up a ramp for bike or skateboard jumps!

* To communicate locations—such as when soldiers need to track an enemy.

* To insert an IV. The IV must go in at a proper angle so the medicine flows correctly. Same thing with self-administered medicine through syringes.

* To sharpen knives. A 20-degree angle is best.

MATH MAGIC TRICK #6

Use this technique to figure out both somebody's age and their shoe size. It's an amazing feet...for the ages!

1 Ask a person to think of their shoe size. (If it's a half-size, have them round up to the next full size.) Have them mentally multiply that number by 5.

Shoe size: 10
$10 \times 5 = 50$

2 Add 50 to that number.

$50 + 50 = 100$

3 Now have the person secretly multiply by 20.

$100 \times 20 = 2{,}000$

4 Add the year it is now, minus 1,000. If it's 2017:

$2{,}000 + 1{,}017 = 3{,}017$

5 Have them subtract the year they were born.

$3{,}017 - 1981 = 1{,}036$

The first two digits are their shoe size. The last two are their age (or the age they'll be this year).

HOW TO REMEMBER MANY DIGITS OF PI

Isn't it impressive when people can recite pi to an extremely high digit? Here's a mnemonic device to help you.

First, an explanation and important comparison: both pi and pie are among the most fascinating things ever to be associated with circles. The former is a mathematical constant—the ratio of a circle's circumference to its diameter, no matter the size of the circle. The latter is made with crust and filling and is delicious. There's one unfortunate difference between the two, however: pi lasts forever; pie does not.

You probably know the first few digits of pi: 3.14159... and so on. It goes on forever—in 2010 a mathematician calculated the constant to its two-quadrillionth digit. You don't need to remember *that* many digits. But a few would be nice. Try a *piem*—that's a pi poem. The number of letters in each word of the piem correlates to a digit.

A piem to remember pi to 15 digits:

How I want a drink, alcoholic, of course, after the heavy lectures involving quantum mechanics!

How (3) I (1) want (4) a (1) drink (5), alcoholic (9), of (2) course (6), after (5) the (3) heavy (5) lectures (8) involving (9) quantum (7) mechanics! (9)

So, pi to 15 digits is 3.14159265358979

Here's a piem to take it to 20 digits:

How I wish I could determine pi
"Eureka!" cried the great inventor.
Christmas pudding, Christmas pie
Is the problem's very center

How (3) I (1) wish (4) I (1) could (5) determine (9) pi (2)
"Eureka!" (6) cried (5) the (3) great (5) inventor. (8)
Christmas (9) pudding (7), Christmas (9) pie (3)
Is (2) the (3) problem's (8) very (4) center (6)

3.14159265358979323846

FIND THE SURFACE AREA OF A PACKAGE

You're going to need to pass the time somehow when you're waiting in line at the post office, and you can't play on your phone because you're holding a huge package (that's why you're at the post office). Might as well figure out how much you're going to be paying when you finally reach the counter.

A box has six sides, so you've got to figure out the surface area of all of them. You could measure each of them individually and add them up, or you could just take three measurements and plug them into this formula.

(Height x Width) x 2
(Height x Length) x 2
+ (Width x Length) x 2
= The surface area of the box

Here's an example.

1 Determine the box's dimensions. (If you don't have a ruler, use a dollar bill—see page 181.) In this example, the box is 12 inches long, 6 inches wide, and 6 inches tall.

2 Plug those numbers into the formula.

$$\begin{array}{r} (6 \times 6) \times 2 \\ (6 \times 12) \times 2 \\ + \underline{(6 \times 12) \times 2} \end{array}$$

$$\begin{array}{l} 36 \times 2 = 72 \\ 72 \times 2 = 144 \\ \underline{72 \times 2 = 144} \end{array}$$

$$72 + 144 + 144 = 360$$

The box has a surface area of 360 square inches. Now you just need to find the newest postage rates for a package that size (the rates seem to increase so often that we can't keep up).

IT'S HIP TO BE SQUARE

Now that you've mastered the skills to do multiplication in your head, you can easily figure out the area of a rectangular surface. This comes in handy when you need to paint a room.

The formula for the area of a rectangular surface is length x width, which is written like this: a = lw.

1 If you have a square wall that's 10 feet by 10 feet, then you simply multiply those two numbers: 10 x 10 = 100 feet^2 (which you would say as "100 square feet"). So you'll need to buy enough paint to cover 100 square feet of a wall.

2 For a rectangle, it's the same formula. If your wall is 10 feet by 15 feet, you multiply length times width, and your area is 150 feet^2.

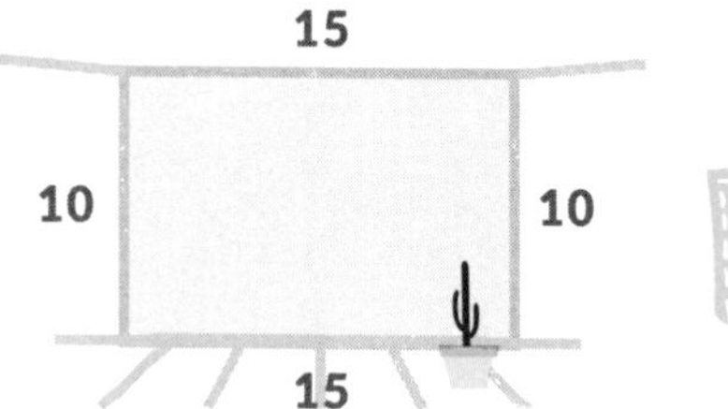

3 Good news! If the shape is a parallelogram, it's the same formula. That's a shape with four sides and each side is the same length as its opposite side, so it looks like a leaning rectangle.

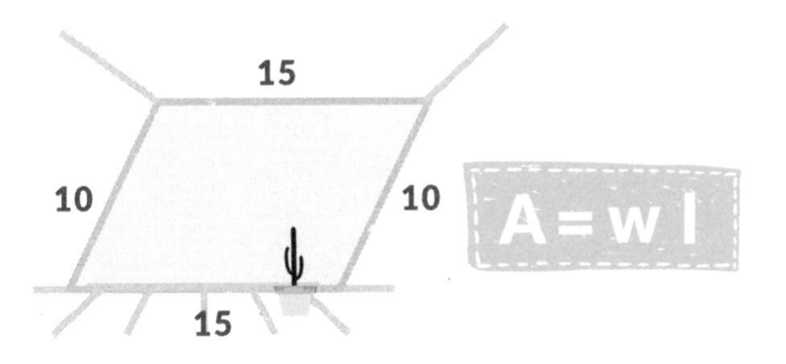

"In mathematics, you don't understand things. You just get used to them."

—Johann von Neumann

SIDE BY SIDE BY SIDE

But our world isn't always divided into simple squares and rectangles. How do you find the area of an oddly shaped room? Let's say the wall you want to paint is on the upper floor of an old house, and the eave is shaped like a triangle.

1. The formula for determining the area of a triangle is:

$$A = \frac{1}{2} bh$$

2. In this case, *b* stands for base, and *h* stands for height. The base is whichever side you determine to be the "bottom." To find the height, measure straight up from the base to the highest point of the triangle.

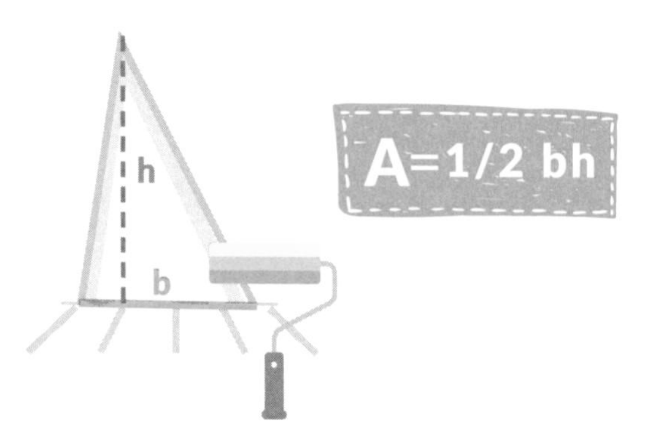

3 So if the triangular section of your wall has a base of 10 feet and a height of 5 feet, then the area will be half of those two numbers multiplied:

$$A = ½ (10 \times 5)$$

$$A = ½ (50) = 25$$

You'll need to buy enough paint to cover 25 square feet of wall.

4 But wait…how much is that? One gallon covers about 350 square feet, so a gallon would be more than enough. If you don't want to buy that much extra, divide your square footage by 350 to get an idea of how many gallons you'll need.

$$25 / 350 = 1/14$$

You need only 1/14 of a gallon, or 0.07. How many ounces is that? Since a gallon has 128 ounces, multiply the number of gallons by 128.

$$0.07 \times 128$$

You get 9 and some change. Remember to round up—buy a pint (16 ounces) to be safe.

MATH MAGIC TRICK #7

This one is also known as the 7-11-13 trick.

1 Ask a friend to write down any three-digit number. Have them tell you what it is.

541

2 Now, have them multiply that number by 7. (This one involves some pretty complex math, so they can use a calculator if they want. They probably haven't read this book. If they had, they'd know the secret of this trick. As for you, you won't need a calculator, because you're up to something.)

541 x 7 = 3,787

3 Have your friend take *that* number and multiply it by 11.

3,787 x 11 = 41,657

4 Finally, your friend must multiply *that* number by 13.

41,657 x 13 =

541,541

5 Tell your friend that the answer is the original number, stated twice.

IT'S A TRAPEZOID!

Oh no! The wall you want to paint is shaped like a trapezoid! What do you do? First, don't panic. The simplest way to find the area of an odd shape is to break it up into smaller, easier shapes.

1 A trapezoid is a four-sided shape with two of its sides parallel. Here, you can easily divide it into a 10' x 10' square and a 10' x 5' triangle. No matter what size the shape is or how many weird angles it has, just break it up into smaller, simpler shapes like rectangles or triangles.

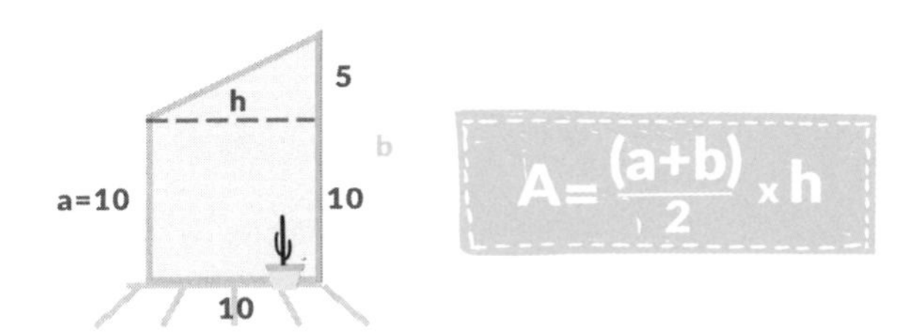

2 Then, it's as easy as adding up the two areas we calculated on the previous pages. The square's area is 100 square feet, and the triangle's is 25 square feet, so your trapezoid has an area of 125 square feet.

You can also use a formula to find the area, where *a* and *b* are opposite sides and *h* is height. In our example, we know the lengths of the left and right sides, so make those sides *a* and *b*. (Think of the trapezoid as lying on its side.) The dotted line is *h*.

$$A = \frac{(a + b) \times h}{2} = \frac{(10 + 15) \times 10}{2} = 125$$

Math-Minded + Famous

- **Supermodel Cindy Crawford won a college scholarship to study chemical engineering (which requires advanced math) but dropped out to pursue modeling.**
- **Before *Desperate Housewives*, Teri Hatcher studied math and engineering in college.**

GOING IN CIRCLES

Here's a curveball: the lower portion of the wall you want to paint is a 10 x 10 square, but the upper portion is a semicircle.

For the upper part, you need to determine the area of a circle. Now you get to use pi! Everyone loves pi! Your friends will be quite impressed when you confidently say, "Hmm, let's see. I have already determined in my head that the area of the square portion is 100 square feet. So to find the area of the semicircle on top, all I need to do is apply the formula $A=\pi r^2$ and then halve it." In plain English:

1 The symbol π refers to pi, a number which is roughly 3.14 (and which all your nerdy friends post about every March 14). And *r* stands for radius. The radius is half of the diameter, the line that cuts a circle in half through the center. If it's easier, measure the diameter first and then divide it by 2 to get the radius.

2 So getting back to our (semi)circle, we've determined that the radius is 5 feet. Plug those numbers into our formula:

$A = (3.14)(5^2)$. Do the easy part first: $5 \times 5 = 25$. Now, using your newly acquired mental math powers (see chapter 1), you can deduce that $3.14 \times 25 = 78.5$.

3 But wait, that's the area of a full circle; you'll need to halve it for a semicircle. And $78.5 / 2 = 39.25$ feet^2.

4 Add that to the 100 square feet you'll need to paint the square portion of the wall, and your total area is (drumroll please)...about 140 square feet.

MAKE WAY FOR SIR CUMFERENCE!

You're not done yet with that wall on page 151—you want to paint a funky orange border that's six inches wide. But unless you're a construction worker or a custom paint shop owner, you don't have a use for leftover orange paint. So how much should you buy?

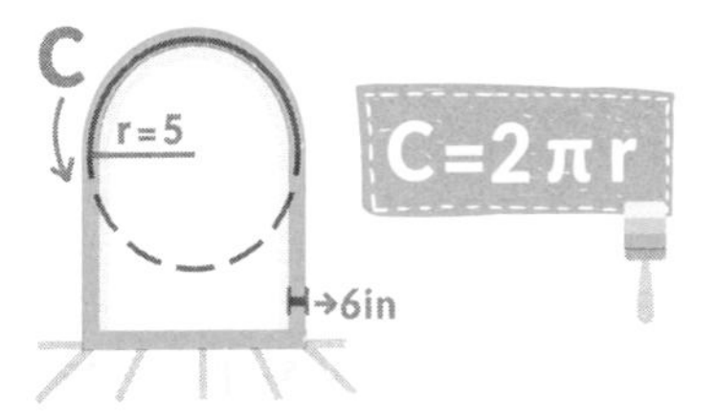

First, let's calculate the length of the border. The 10'x 10' portion will require a line spanning the floor and the two walls. That's 10 + 10 + 10 = 30 feet. But how many feet do you paint along the semicircle arch at the top? To determine that, you need to find the circumference, or the distance around the circle.

1 The formula for circumference is $2\pi r$. Use 3.14 for π:

$$(2)(3.14)(5) = 31.4$$

Or if it's easier, then you can use this formula: πd, where *d* is the diameter. The diameter is twice the length of the radius.

2 The circumference of the full circle is 31.4 feet, so to paint around your semicircle, you'll need to paint half that, or about 16 feet. Add that to the total distance along the floor and up the two walls, and your border will be about 46 feet.

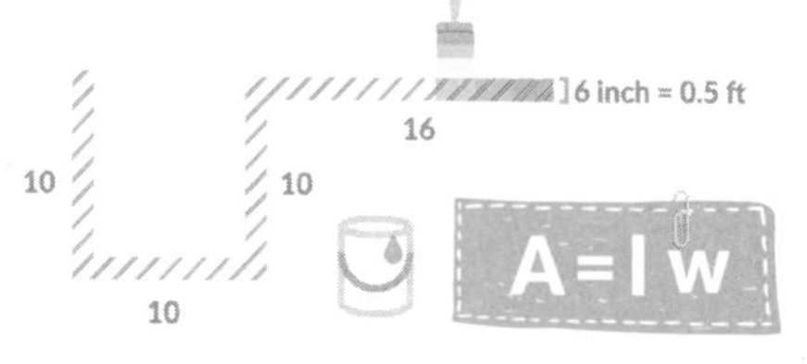

3 Now you can calculate the area of your border, which is basically a rectangle that's 46 feet long and 6 inches wide. To plug that into our area formula, you'll need both numbers to be expressed in the same unit of measurement—either both in inches or both in feet. Simply change six inches to 0.5 feet, and your formula looks like this: A = (46) (0.5). You'll need to buy enough orange paint to cover 23 square feet. Since one gallon of paint covers about 350 square feet, that equals only about 0.07 gallons. A pint should do it.

Speaking of pints, you're going to throw a party in your newly painted house! Go to page 159 to learn how to use math to stock up on liquid refreshments.

PYTHAGOREAN THEOREM

In the sixth century B.C., Greek mathematician Pythagoras unlocked a secret about triangles...and squares.

If a triangle has a right angle, and you build a square off of each of the three sides, then the area of the largest square (built off the triangle's longest side) will have the same area as the other two squares combined.

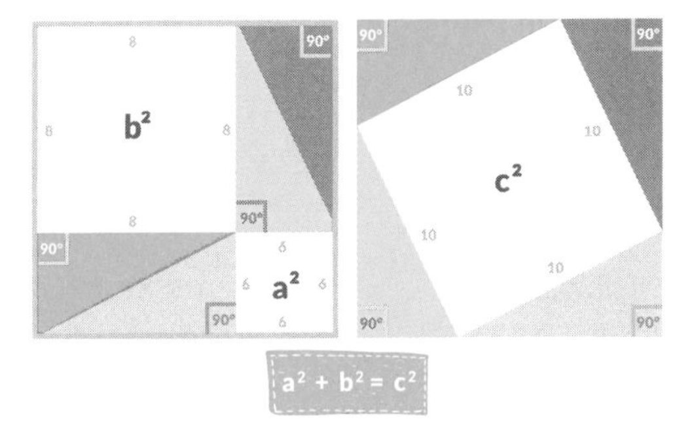

It looks like this if you express it as an equation: a, b, and c are the sides of the triangle, and *c* is the largest side, also called the hypotenuse:

$$a^2 + b^2 = c^2$$

Say a triangle has sides that measure 6 inches, 8 inches, and 10 inches.

* The square made from the 6-inch side would have four sides consisting of 6 inches each. The area of a square is determined by multiplying the length of two sides together (which, in a square, are the same). So the area is 6^2, or 36.

* The square made from the 8-inch side has an area of 8^2, or 64.

* The square made from the hypotenuse is 10^2, or 100.

$$a^2 + b^2 = c^2$$
$$6^2 + 8^2 = 10^2$$
$$36 + 64 = 100$$

The Pythagorean theorem can also be used to find the length of an unknown side of a triangle if the other two sides are known. Just plug in the values and solve like an algebraic equation.

$$a^2 + b^2 = c^2$$
$$5^2 + 12^2 = c^2$$
$$25 + 144 = c^2$$
$$169 = c^2$$
$$c = \sqrt{169}$$
$$c = 13$$

MATH MAGIC TRICK #8

There are easier ways to get somebody's digits, but this trick is still worth a shot. It starts with part of a phone number...and ends with the whole thing.

1 Take the first three digits of your phone number—not including the area code.

867 – 5309
867

2 Multiply by 80.

$$\begin{array}{r} 867 \\ \times\, 80 \\ \hline 69{,}360 \end{array}$$

3 Add 1.

$$\begin{array}{r} 69{,}360 \\ +\, 1 \\ \hline 69{,}361 \end{array}$$

4 Multiply by 250.

$$\begin{array}{r} 69{,}361 \\ \times\, 250 \\ \hline 17{,}340{,}250 \end{array}$$

5 Add the last four digits of your phone number.

$$\begin{array}{r} 17{,}340{,}250 \\ +\ 5309 \\ \hline 17{,}345{,}559 \end{array}$$

6 Add the last four digits of your phone number again.

$$\begin{array}{r} 17{,}345{,}559 \\ +\ 5309 \\ \hline 17{,}350{,}868 \end{array}$$

7 Subtract 250.

$$\begin{array}{r} 17{,}350{,}868 \\ -\ 250 \\ \hline 17{,}350{,}618 \end{array}$$

8 Divide by 2.

$$\begin{array}{r} 17{,}350{,}618 \\ \div\ 2 \\ \hline 8{,}675{,}309 \end{array}$$

8,675,309

867 – 5309

The answer is your entire phone number!

PUMP UP THE VOLUME

It's party time! You have a big metal drum that's been waiting years for a batch of fruit punch, but how much juice (and other, more spirited ingredients) will it fit? For that, you'll need to determine the volume of your metal drum, which in math terms is called a cylinder. (You'll want to write this one out.)

1 A cylinder has a circle as its base. Finding the volume of a cylinder is similar to finding the area of a circle, except now you're dealing with three dimensions. So here's the formula:

$$V = bh$$

2 The *b* stands for the area of the base, which is a circle. Its area is: $A=\pi r^2$. Let's plug that into our formula:

$$V = \pi r^2 h$$

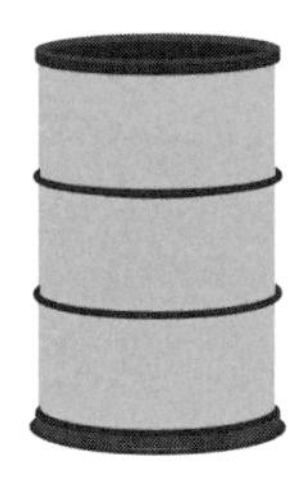

3 The *h* stands for the height of the drum. As usual, make sure all numbers are in the same unit of measurement. Say your drum has a diameter of 18 inches (which is a radius of 9 inches) and is 3 feet high. Convert the height to 36 inches, and your formula will look like this:

$$V = (3.14)\ (9^2)\ (36)$$
$$V = (3.14)\ (81)\ (36)$$

Now multiply 3.14 and 81, and then multiply that figure by 36, and your volume is 9,156 inches^3 (cubic inches).

4 But if you go to the store and say, "I'd like to buy 9,156 cubic inches of punch ingredients," then you'll get a funny look from the clerk. You'll need to change that to U.S. gallons. There are 231 cubic inches in a gallon. So divide 9,156 by 231 (good luck!), and you can tell the confused grocery store clerk, "I'd like to buy 39.63636 gallons of punch ingredients."

The largest prime number discovered as of press time is more than 22 million digits long. It was found using GIMPS...software that stands for the Great Internet Mersenne Prime Search.

CHAPTER WRAP-UP

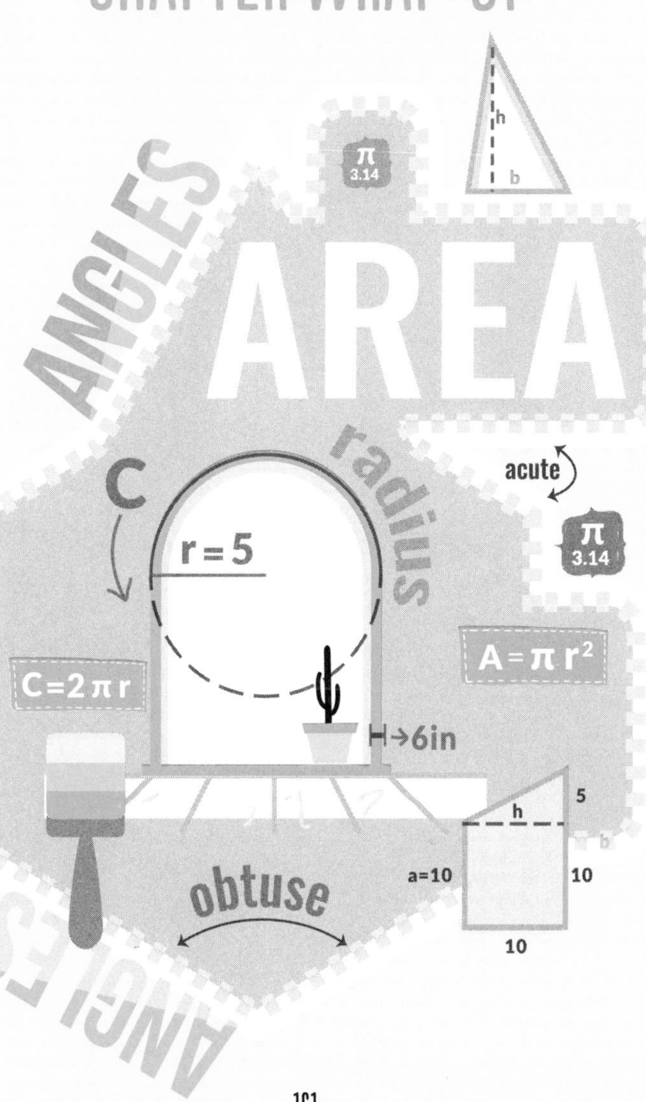

5. MATH IN ACTION

USE YOUR WATCH AS A COMPASS

LEARN CLEVER MATH MNEMONICS

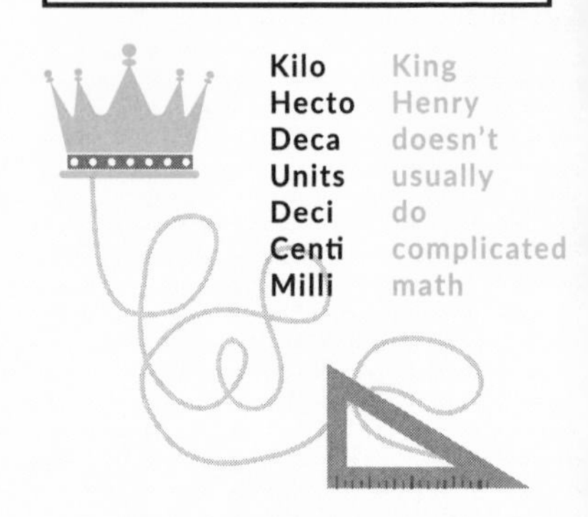

Kilo King
Hecto Henry
Deca doesn't
Units usually
Deci do
Centi complicated
Milli math

HAVE FUN WITH MATH GAMES

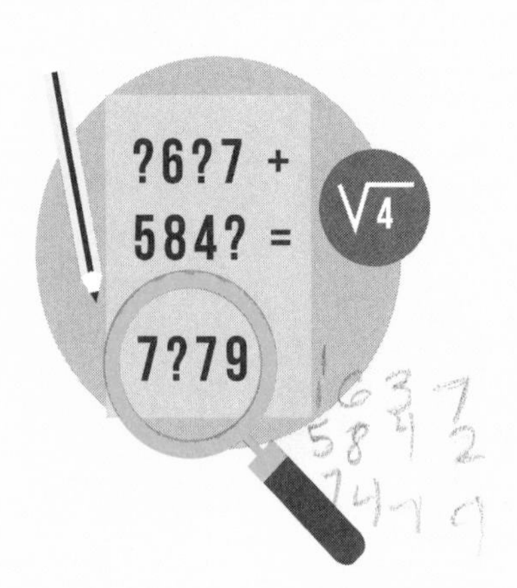

MAKE A MATH CLOCK AND OTHER PROJECTS

MATH MNEMONICS TO HELP YOU OUT

This book is full of easy tricks to remember math operations. But the mnemonic devices—little sayings or rhythmic turns of phrase—here are even simpler.

How to... Find the area of a circle

Mnemonic device: (Forgive the poor grammar.)
"Pies are round, but pi r squared."

Math operation:
The area of a circle = $\pi \times r^2$

Note: r = radius (which is half the diameter)

Example:

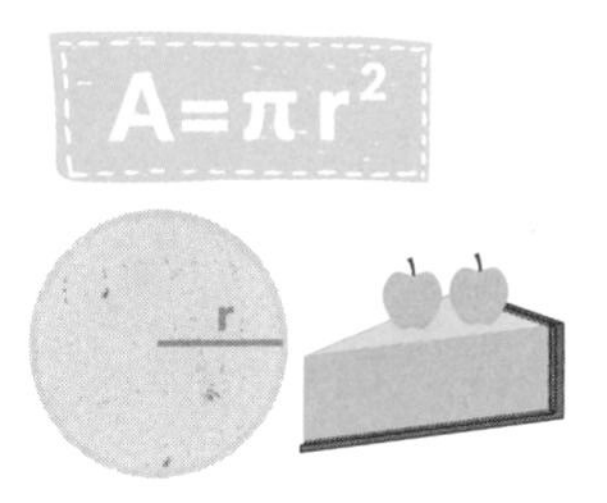

How to… Find the circumference of a circle

Mnemonic device: (Let's also use a pi/pie homonym here.) **C**herry **p**ie **d**elicious.

Math operation: C = p (for π) x d

Note: Circumference = π x diameter

Example:

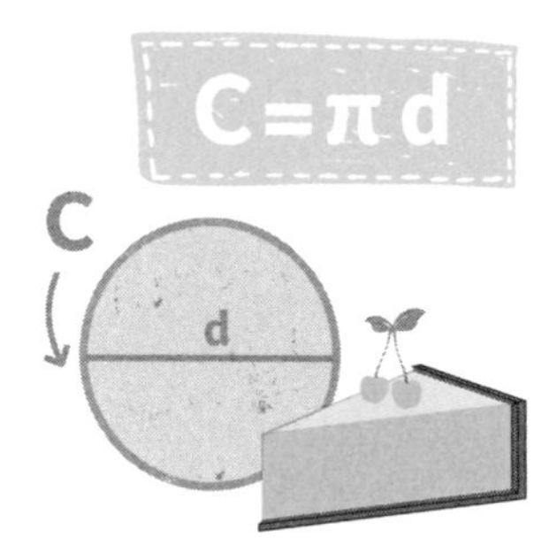

How to… Divide fractions

Mnemonic device: Can **F**red **m**ultiply?

Math operation:

1 **C**hange the division to multiplication.

2 **F**lip the second fraction.

3 **M**ultiply.

Example:

$$\frac{3}{4} \text{ divided by } \frac{1}{4}$$

$$1.\ \frac{3}{4} \times \frac{1}{4}$$

$$2.\ \frac{3}{4} \times \frac{4}{1}$$

$$3.\ \frac{3 \times 4}{4 \times 1} = \frac{12}{4} = 3$$

How to... Tell the difference between a numerator and a denominator

Mnemonic device: There are two numbers in a fraction: one is above the line, and the other is below the line. The *numerator* is above the line, and the *denominator* is below. To stay on top of this, remember the phrase, "nice dog."

Explanation: Nice comes *before* dog, or N (numerator) before D (denominator).

ALL HAIL ROMAN NUMERALS!

Can't remember if you've seen this Rocky movie or which Super Bowl we're on? Never fear! You can figure it out using this easy trick to translate Roman numerals into regular numbers and back again.

Roman numerals are depicted by letters of the alphabet. Only seven letters are used: I, V, X, L, C, D, and M. Each is introduced as numbers get higher, but it's hard to remember in what order. Try this:

I **V**IEWED **X**ENA **L**OPING **C**ARELESSLY **D**OWN **M**OUNTAINS.

I = 1
V = 5
X = 10
L = 50
C = 100
D = 500
M = 1,000

But remember: the Roman numeral system involves not

just adding more letters, but subtracting them when the number is approaching a new letter. For example, 80 is LXXX (50 plus 10 plus 10 plus 10), but 90 is XC (100 minus 10).

Now, let's try converting some Roman numerals.

Q What number Super Bowl would Super Bowl LXXV be?

L = 50
X = 10
X = 10
V = 5

A It's Super Bowl 75.

Q What overall installment is *Friday the 13th Part CCXLII?*

C = 100
C = 100
X = 10
L = 50
II = 2

A That adds up to 262. But wait! There's an XL in there—which means the 10 is subtracted from the 50, which gives you 40. That means you're waiting in line to see *Friday the 13th Part 242.*

STAY POSITIVE, OR GO NEGATIVE?

When multiplying negative numbers, either with other negatives or positives, it can get confusing remembering whether to turn the solution negative or positive. Here are some ways to help you remember.

Mnemonic device: A good thing that happens to a good person is good.

It means that... A positive number multiplied by a positive number equals a positive solution.

Example:

$$\begin{array}{r} 8 \\ \underline{\times 4} \\ 32 \end{array}$$

Mnemonic device: A good thing that happens to a bad person is bad.

It means that... A positive number multiplied by a negative number equals a negative solution.

Example:

$$\begin{array}{r} 8 \\ \times\ -4 \\ \hline -32 \end{array}$$

Mnemonic device: A bad thing that happens to a good person is bad.

It means that… A negative number multiplied by a positive number equals a negative solution.

Example:

$$\begin{array}{r} -8 \\ \times\ 4 \\ \hline -32 \end{array}$$

Mnemonic device: A bad thing that happens to a bad person is good.

It means that… A negative number multiplied by a negative number equals a positive solution.

Example:

$$\begin{array}{r} -8 \\ \times\ -4 \\ \hline 32 \end{array}$$

LESSER OR GREATER?

The < and > symbols are used to indicate when a number is "less than" or "greater than" another number. The < means that the number on the left is smaller than the one on the right. The > means that the number on the left is larger than one on the right. Here are little ways to help you remember which is which.

★ **The Alligator or Pac-Man Method.** The < looks like an alligator, right? Well, the alligator is a hungry beast, and it would "eat" the larger number. Similarly, Pac-Man would gobble up the larger of the two numbers.

★ **The Bigger Side Method.** The right-hand, open side of the < is the bigger end. And the bigger end faces the bigger number.

★ **The German Method.** *Kleiner dan* is German for "smaller than." In some German-speaking areas of Europe (such as Belgium), math students add a vertical line to their > or < signs. Like so: |> and |<. The one that means less than looks almost like a *k*...the first letter of *kleiner dan*, which means "smaller than."

MORE MNEMONICS TO HELP YOU OUT

Here are a few more mnemonic devices to advance your math skills.

How to... Tell the difference between mean, median, and mode.

Mnemonic device: They all *kind* of mean the average, but not really. Here's how to tell them apart:

Mode is the most frequent occurrence.

Median is the middle—like how roads have medians in the middle. Line up the numbers in order, then find the middle one.

Mean is the average—because it's pretty average to be mean.

Example: Let's look at the different "averages" of one set of numbers.

1 1 1 2 3 5 8

Mode

1 1 1 2 3 5 8

The mode is 1.

Median

1 1 1 **2** 3 5 8

The median is 2.

Mean

$$1 + 1 + 1 + 2 + 3 + 5 + 8 = 21$$

$$21 / 7 = 3$$

The mean is 3.

How to... Remember the metric units of measurement.

Mnemonic device: King **H**enry **d**oesn't **u**sually **d**o **c**omplicated **m**ath.

Explanation:

Kilo

Hecto

Deca

Units

Deci

Centi

Milli

How to... Define an isosceles triangle.

Mnemonic device: An isosceles triangle is one in which two sides and angles are equal. To remember that, sing this phrase to the tune of "O Christmas Tree."

"Isosceles, Isosceles,
Two angles have equal degrees.
Isosceles, Isosceles,
You look just like a Christmas tree."

Example:

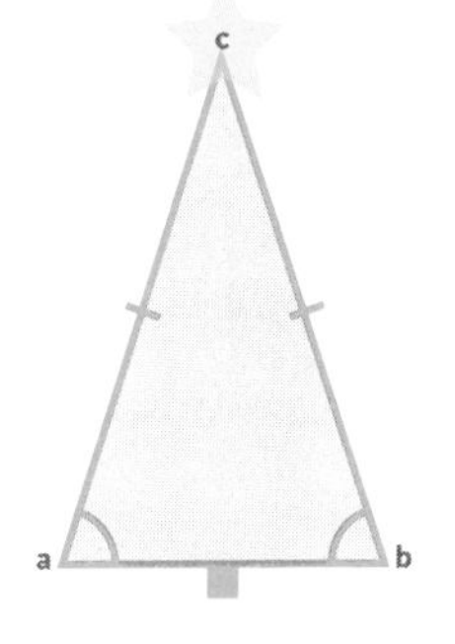

MATH MAGIC TRICK #9

This one requires a calculator, pencil, and paper.

1 Write "73" on a piece of paper, fold it up, and give it to an unsuspecting victim...er, person. Tell her she's not to look at it until you say so.

2 Ask her to think of a four-digit number and enter it twice into a calculator, for example: 36,243,624.

3 Tell her to divide the number by 137.

4 Have her divide that result by the original four-digit number.

5 Tell her to open the piece of paper you gave her earlier. It will match the display on the calculator. That's right—the amazing number 73!

Want to know how it works? Entering a four-digit number twice, such as 36,243,624, is the same thing as multiplying it by 10,001. (3,624 x 10,001 = 36,243,624). Because 10,001 = 73 x 137, the eight-digit number will be divisible by 73, 137, and the original four-digit number.

7 MEASURING HACKS

What if you need to measure something, but for some reason you forgot to put your ruler or your tape measure in your back pocket? No worries. You can use regular objects you've got on you.

1. A standard credit card is about 3.5 inches wide and 2 inches long.

2. Don't carry plastic cards? Grab a pen. A standard disposable pen is 6 inches long.

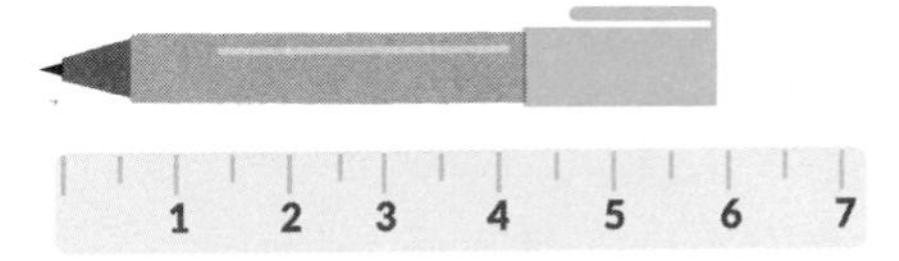

3. Need an inch? Use a quarter. Its width is 0.955 inches...so almost an inch.

4. Need another inch? Your house key is just about 2 inches long.

5 A standard square makeup compact has sides that measure right around 2.5 inches.

6 Or you could measure your hand, or the distance from your elbow to your wrist, or the length of your index finger. Memorize them, and you'll always have a ruler (or two, or three) on your person.

7 Hold your arms out straight at your sides. The distance from fingertip to fingertip usually equals your height—another good one to memorize and use.

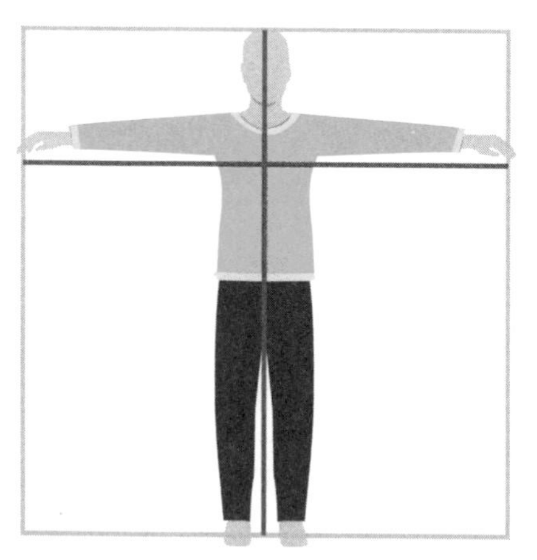

8 Hold one arm out straight again, with your head facing forward. The distance from your nose to the tip of your middle finger is about 3 feet (unless you have freakishly long or short arms).

MEASURE WITH A MIRROR

Want to know how tall that building over there is? Grab a mirror and a yardstick.

1 Place the mirror on the ground, faceup. Start walking away from it, backward, toward the building (or flagpole, or whatever). When you can still see the mirror, and you can see the very top of the building reflected in the mirror, stop walking.

2 Mark on the ground where you stopped. Now, using the yardstick, measure the distance from the stop spot back to the mirror.

Distance #1 = 10 feet

3 Next, measure the distance from the mirror all the way to the base of the object being measured.

Distance #2 = 30 feet

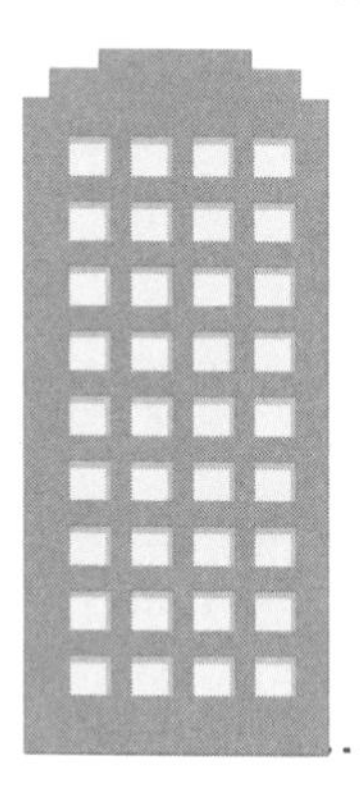

4 Multiply distance #2 by the height of your eyeballs from the ground. (You don't have to measure yourself—just take your height and subtract two inches.)

5 feet, 2 inches – 2 inches = 5 feet
30 feet x 5 feet = 150 feet

5 Divide that answer by distance #1. The result is the height of the object.

150 feet / 10 feet = 15 feet

FOOD MEASURING TRICKS

One of the most common places you'll have to measure things is the kitchen. But what if you're following a recipe and the tablespoon is dirty, or you don't have a 1/3-cup measuring cup, or a scale? Here are some measurement hacks and equivalencies.

* A standard spoon in a silverware set is a teaspoon.

* 3 teaspoons is equal to 1 tablespoon.

* 24 teaspoons is equal to 1/2 cup.

* 2 tablespoons is the same as 1/8 cup, and 16 tablespoons = 1 cup.

* 1/3 cup is equivalent to 5 tablespoons plus 1 teaspoon.

* You probably know that there are 8 ounces in a cup. Moving up the liquid measurement ladder, 2 cups equals 1 pint, 2 pints equals 1 quart, and 4 quarts equals 1 gallon.

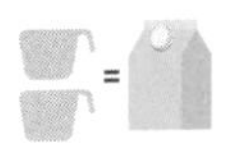

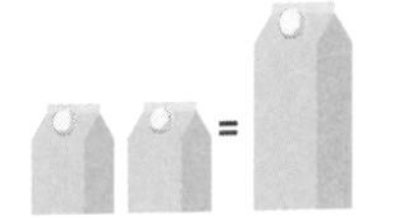

* A serving of chicken or beef is 3 ounces—or about the size of a deck of cards. Fish is a bit flatter and thinner than other meats. A 3-ounce portion of fish is about the size of a checkbook.

* When using fats, think of poker chips. The width and thickness of a chip is about the same size as 1 ounce of butter, salad dressing, mayonnaise, or cooking oil.

=

1 cup

* One cup's worth of dry goods—cereal, rice, or flour, for example—is roughly the size of a baseball. So is a cup of salad greens or a cup of cooked vegetables.

* Two tablespoons of peanut butter takes up the same amount of space as a golf ball.

MEASURE WITH A DOLLAR BILL

You're at the store and need to measure something to make sure it'll fit in your home, or your car. Well, you probably have at least a dollar bill on you, right? Here's how to turn it into a ruler.

Perhaps for the first time ever, it doesn't matter whether you have a dollar or a hundred dollar bill—it's all the same for this trick. Any American bill is 6.25 inches long and 2.5 inches wide. Measure the object in question by holding the bill up lengthwise or widthwise as many times as needed to move all the way across it. It's that simple!

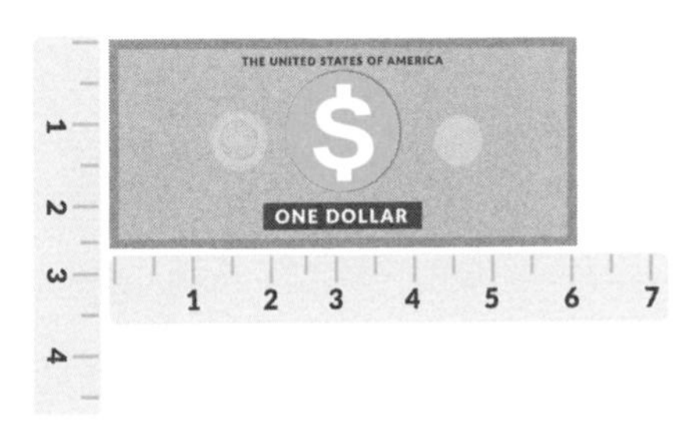

Here's how to estimate size:

* The full length of the bill is about 6 inches. So, two bills across is about one foot.
* Fold the bill in half to estimate 3 inches. Or measure without folding—3 inches is the distance from either side to the "o" in "dollar."
* To measure something even smaller, fold it in half again; that's 1.5 inches. That's the distance from the left side to the center of the Federal Reserve seal (it has a letter that signifies where the bill was printed).
* Using the width of the bill, you can measure in 2.5-inch increments. Twice its width is 5 inches. Half its width is about 1 inch.

If the object is one bill long plus two bills wide, it's

$$6 + 2.5 + 2.5 = 11 \text{ inches.}$$

So go forth and stretch the power of your dollar!

MEASURE A REALLY TALL THING WITH A STICK

Ever want to know how tall a building or tree is? And you don't have a crane and a giant tape measure? Just use a stick.

What You'll Need:

A really tall object

A stick

A ruler (optional)

What to Do:

1. Find a stick (or another object of uniform length) that's the size of your arm from shoulder to fingertips.

2. Hold the bottom of the stick in your hand so it's pointing straight up, and hold your arm out straight so that the stick and your arm form a 90-degree angle.

3 Now, making sure not to hit anything, walk backward while focusing your vision on the top of the stick. Stop walking when the top of the stick lines up with the top of the thing you're measuring.

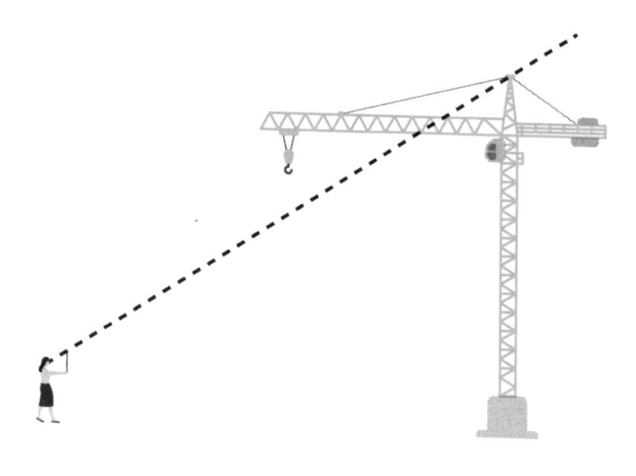

4 Provided that the ground is level, you're now about the same distance from the tall thing as the tall thing's height.

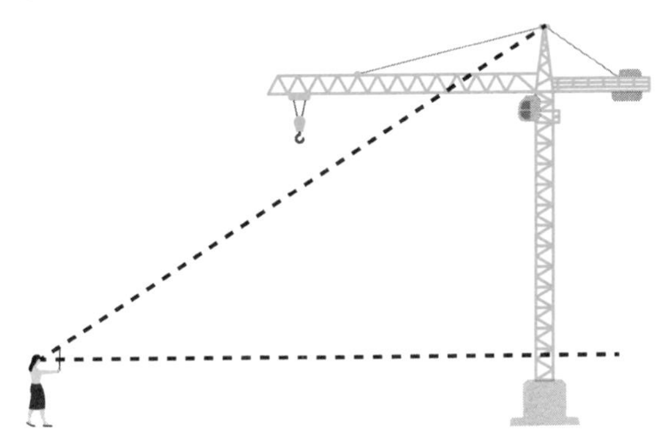

5 Measure (or estimate) the length of your foot, and walk back to where you started, putting one foot carefully in front of the other with no space in between. (Estimation tip: The average woman's foot size is 9.5 inches; for men it's 11 inches.)

6 Multiply the length of your foot by the number of steps you took. That will give you the height of the tall building or tree.

WRISTWATCH COMPASS

Until you download that compass app for your smartphone, here's a simple trick to help you find your way around the outdoors.

What You'll Need:

An analog watch (the kind with hour and minute hands) set to the correct time

A pen, pencil, or twig

A clear view of the sun

What to Do:

Poke the pen upright into the ground so that it casts a shadow.

Set the watch on the ground in the shadow cast by the pen.

Rotate the watch until the hour hand lines up with the shadow and points at the sun. (The shadow is cast directly away from the sun, so aligning the hour hand with it also aligns it with the sun.)

Note the angle formed by the hour hand and the 12 o'clock mark on the face of the watch. Now divide it in half to find the midpoint. Example: if it's 8:00, the hour hand points to the 8. The midpoint between the 8 and the 12 is the 10.

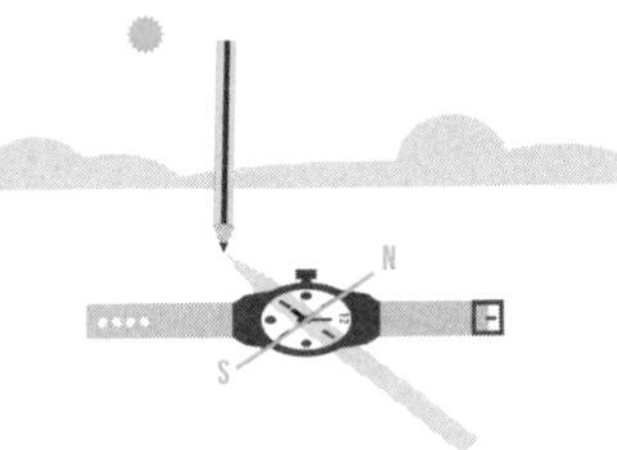

The line created by the midpoint number (10 in this example) and the number opposite it on the watch face (4) form a north-south line. If you're not sure which is which, remember that the sun is in the east before noon and in the west after noon. If the 10 points north and the 4 south, then east (and the sun) is at 1, and west is at 7.

Tips:

Daylight Saving Time? From March to November, your watch is set an hour ahead, so find the north-south line by bisecting the angle created by the hour hand and 1 o'clock instead of 12 o'clock.

South of the equator? Point the 12 at the sun instead of the hour hand and bisect the angle they create to find the north-south line.

Wearing a digital watch? A carefully drawn picture of a clock with the hour hand pointing to the correct time works just as well.

PYTHAGOREAN IN ACTION

The Pythagorean theorem is a nifty way to figure out the length of one side of a right triangle if you know the lengths of the other two sides. Perhaps you want to know how long a ladder needs to be to reach a certain height on a wall...or how big a TV will fit into the corner of a room.

The theorem applies only to right triangles, which have a 90-degree angle. It tells us that when you measure the two sides adjacent to the right angle, square their lengths, and add them together, the result will equal the square of the length of the diagonal side. (For more about this theorem, see page 155) The formula is: $a^2 + b^2 = c^2$. Demonstrate it for yourself.

What You'll Need:

Three paper clips

Three straws

A pair of scissors

A ruler

What to Do:

1 Bend one of the paper clips into a right angle.

2 Use the scissors to cut two of the straws into different lengths. Let's say three inches and four inches.

3 Push one end of the right-angled paper clip into the three-inch straw—side *a*—and the other end into the four-inch straw: side *b*.

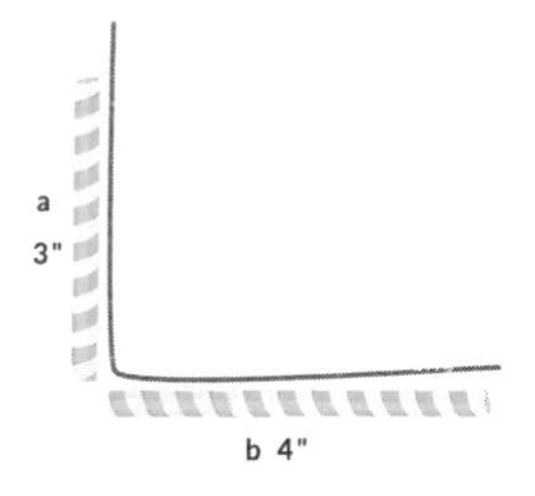

4 Take the third straw and use it to make a diagonal with the other two straws. Cut it to fit the diagonal's length. This is side *c*.

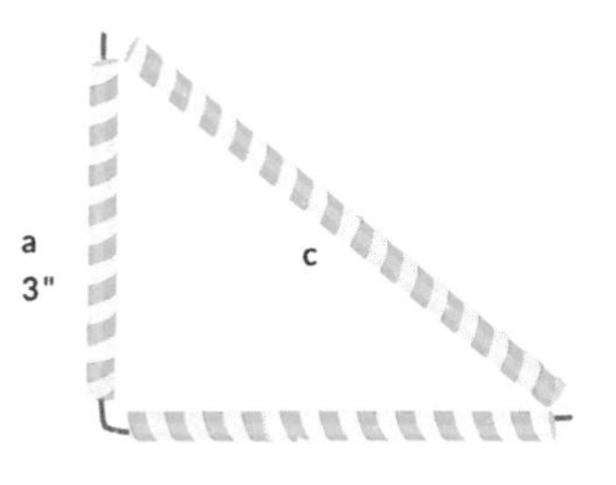

5 Bend the other two paper clips and use them to attach the ends of side *c* to the open ends of side *a* and side *b*. All three straws together now form a right triangle.

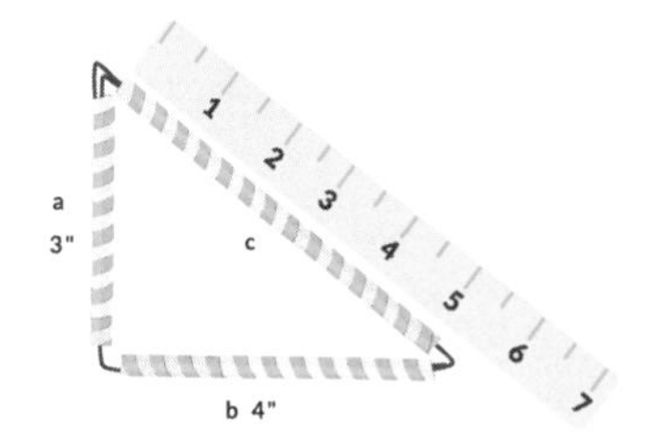

6 Now apply the formula with side *a* as 3 inches and side *b* as 4 inches:

$$a^2 + b^2 = c^2$$

$$3^2 + 4^2 = c^2$$

$$9 + 16 = c^2$$

$$25 = c^2$$

$$\sqrt{25} = 5$$

The third straw should be 5 inches long. Measure it and see.

MAKE A SUNDIAL

This sundial will give you an approximate time, but don't rely on it to keep appointments!

Some of the earliest known sundials date back to 1500 B.C., long before clocks and watches. Through modern-day knowledge of the Earth's movements (and that Earth is actually round!), we now know that they require sophisticated math—since a summer day in Alaska is longer than the same day in Dallas, a sundial's angles must be adjusted based on location in order to be accurate.

What You'll Need:

Paper plate

Marker

Straw

Tape or glue

Protractor

Sunshine!

What to Do:

1 Use your protractor to mark and draw straight lines from the center of the plate out to the edge, in equal measurements of 30 degrees. (Why 30? A clock has 12 numbers, and there are 360 degrees in a circle. 360÷12= 30. Each mark represents a number on the clock.)

2 Poke a hole in the center of the plate and guide the straw through. Reinforce the straw in an upright position with tape underneath or by filling the center hole with glue or museum putty. This creates the gnomon, or clock hand, of the sundial.

It's pronounced
'GNOME-mon'.

3 Place it squarely on the ground outside. Align the straw's shadow with one of the lines you drew with the protractor. Mark it with the current time. (You'll get the best results in the early afternoon and on the hour, around 1 p.m. or 2 p.m.)

4 Leave the sundial on the ground and return to it in an hour. Where is the shadow now? What time must it be? Mark the next line accordingly. Based on the direction the shadow moved, sequentially label the rest of the lines you drew. This gives you the approximate hour markings for your location.

Now if you ever want to know the time during, say, a power outage, you can go outside and see where the straw's shadow falls!

Tip:

The angles for the sundial's hour markers and the angle and size of the gnomon need to be customized to each region. You can search online using keywords like "sundial gnomon angle calculator" for more specific calculations.

A pizza that has radius ***z*** and height ***a*** has volume Pi × z × z × a.

DIVIDE ANYTHING INTO EQUAL PARTS

Here's a neat trick to split up any straight surface into equal sections...without having to deal with fractions!

Carpenters and other workers use this shortcut to divide paper, wood, or cabinets without having to measure them. It's especially useful if the object is an odd measurement that doesn't divide evenly.

1 It works like this: say you have a rectangular piece of wood that's 1 foot 11 $\frac{1}{2}$ inches across. First, determine how many sections you want to divide it into. Let's say five equal parts.

2 With a ruler or measuring tape, measure a diagonal line that goes above the side you want to split up. The line should form an angle less than 90 degrees, and its length should be easily divisible by the number of sections. In our example, let's position a 50-inch-long line from the lower left side up to the right side.

3 Now mark equal measurements along the diagonal line. In this case, five sections of 10 inches means you mark 10, 20, 30, and 40 inches. If the line was 30 inches, you'd mark 6, 12, 18, and 24 inches instead.

4 Finally, use a straight edge to draw lines down from each mark to the bottom edge of the object. It's now split up into equal sections!

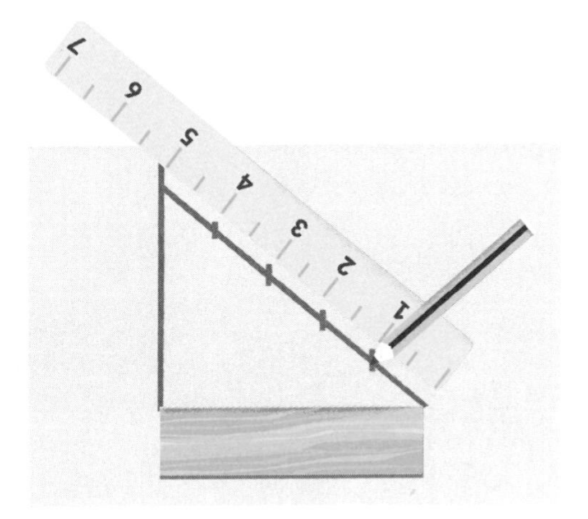

1 If the object is too small for you to draw a diagonal line above it—if it's a one-inch-high piece of wood, for example—then start by placing it along the bottom edge of a piece of paper or something else you can draw on. Draw a vertical line up from one side of the wood. Now position the ruler or measuring tape diagonally above the wood to a point on the vertical line you drew. Make sure the diagonal measures a number that's divisible by the number of sections. Then follow the rest of the instructions.

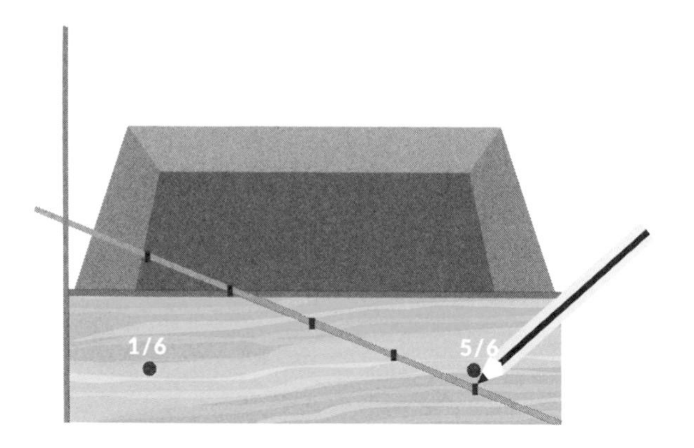

This trick works whether you'd like to divide a paper into seven columns to make a calendar, or center knobs at 1/6 and 5/6 of the way across a bureau drawer. You'd divide them without even figuring the exact measurements—easy peasy!

MATH MAGIC TRICK #10

All you need to know is the date, and you can find the day of the week that pretty much anything happened...as long as it was in the last century.

1 First you need a code for each of the 12 months. These are a bit tough to memorize, so we've included some memory tricks to help you out.

January	6 (There are six letters in "winter.")
February	2 (February is the second month of the year.)
March	2 ("March 2 the beat of your own drum!")
April	5 (April has 5 letters.)
May	0 (How much school or work on May's Memorial Day? 0.)
June	3 (Consider the june b-u-g—3 letters.)
July	5 (It's the month you set off "fiver-works"!)
August	1 (The first letter of August is A, or the alphabet's first letter.)

September	4 (It's the first month of F-A-L-L.)
October	6 (There are 6 letters in "Tricks" and "Treats.")
November	2 (At Thanksgiving you eat 2 times as much as you normally would.)
December	4 (There are 4 letters in XMAS.)

2 Now you need a number that correlates to the year of the date in question. First, get a code for the decade of the year being guessed.

1900s	1	**1940s**	2	**1980s**	3
1910s	6	**1950s**	0	**1990s**	1
1920s	5	**1960s**	6	**2000s**	0
1930s	3	**1970s**	4	**2010s**	6

3 The last part is a number for the year that offsets how Leap Year makes some years have an extra day. Decades are either even-numbered (like the 1980s) or odd-numbered (the 1930s).

	Even decade	Odd decade
Year ends in 0	0	0
Year ends in 1	0	0
Year ends in 2	0	1
Year ends in 3	0	1
Year ends in 4	1	1
Year ends in 5	1	1
Year ends in 6	1	2
Year ends in 7	1	2
Year ends in 8	2	2
Year ends in 9	2	2

Let's try it. Say your cousin had her baby on April 21, 2006.

1 Take the day of the month, and add to it the month code from the chart above.

April 21
21 + 5 = 26

2 Add to that figure the decade code and the last digit of the year.

2000s - 0
Year of birth - 2006 - 6

$$26 + 0 + 6 = 32$$

3 Add the leap year offset.

2006 is an even-numbered decade that ends in 6: +1

$$32 + 1 = 33$$

4 Divide the figure by 7 (because there are 7 days in a week). Discard the solution and keep only the remainder.

$$33 / 7 = 4, \text{remainder } 5$$

5 Take the remainder and match it up with the day of the week on the chart below. (These aren't hard to remember, if you consider Monday to be the first day of the week, or if you think about how Tuesday sounds like Two's Day.

$$\text{Friday} = 5$$

Your cousin's baby was born on April 21, 2006—a Friday. Now you just need a trick to remember birthdays!

Monday	1	**Friday**	5
Tuesday	2	**Saturday**	6
Wednesday	3	**Sunday**	0
Thursday	4		

MAKE A MATH CLOCK

Here's a fun way to practice math every time you look at the time!

What You'll Need:

A manual (analog) clock

A sheet of construction paper

A pencil and eraser

Scissors

A protractor

A marker

Glue

What to Do:

1 Lay the paper or poster board over the clock face. Using a pencil, trace the clock face. You might want to make it just a bit bigger than actual size.

2 Use scissors to cut the paper.

3 Mark the center and the top 12 o'clock spot with a pencil.

4 Position the protractor at the center, measure 30 degrees to the right, and make a mark. Lightly draw a line from that to the outer edge of the circle. That's 1 o'clock.

5 Mark the other hours on the clock in the same way, with each hour in 30-degree increments.

6 Now, instead of writing the numbers for each hour, use the marker to write a math problem whose answer is the number. It can be as simple or as complex as you like. For 1 o'clock, you could write:

$$\frac{3}{4} \times \sqrt{\frac{16}{3}}$$

For 9 o'clock, one possibility is $(\pi - .14)^2$.

7 Erase all the pencil marks.

8 Cut a line up from the bottom of the circle to its center. Cut a hole in the center big enough to fit around the clock's moving hands.

9 Position the paper so the hands move in front of it, and glue it to the existing clock face.

MAKE A PROTRACTOR

Here's a fun way to make a protractor simply by folding up a sheet of paper.

What You'll Need:

A sheet of $8\frac{1}{2}$ x 11-inch printer paper

A ruler

A pencil or pen

A pair of scissors

What to Do:

1 With the ruler, measure $8\frac{1}{2}$ inches down the longer side of the paper and make a mark. Then draw a line from the mark straight across the paper. With the scissors, cut the paper along the line. The sheet of paper should now be a square: $8\frac{1}{2}$ inches long on all four sides.

2 Fold the paper in half and unfold it.

3 Fold the upper right corner of the paper to touch the corner to a low point about two-thirds of the way down in the crease you made in step 2. The folded part will now be a triangle with 30-, 60-, and 90-degree angles.

4 Fold the bottom right corner of the sheet up to the edge of the triangle you just made to form a second triangle.

5 Fold the bottom left corner of the sheet up until it meets the edge of the first triangle. Tuck this third triangle under the second triangle (the one you made in step 4). Your paper should now be shaped like a large triangle consisting of three smaller interlocking triangles.

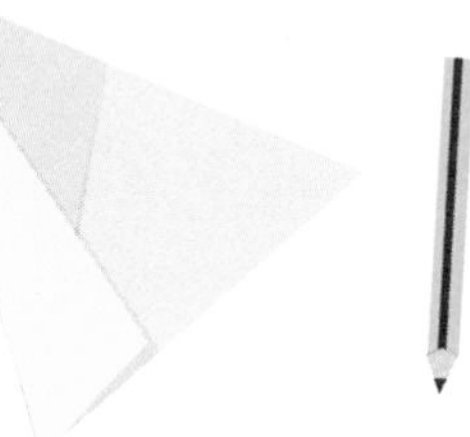

6 You have a makeshift protractor! Now label all the angles. Lay the paper down with the longest side pointing up. At the top corners are two angles separated by a line. Label the left one as 15° and the right one as 30°.

7 The left corner also has two angles on either side of a line. Mark the one on the left as 45°. The one on the right is 30°.

8 Label the right corner as 60°.

9 On the right side of the protractor is a line between two 90° angles. Label these.

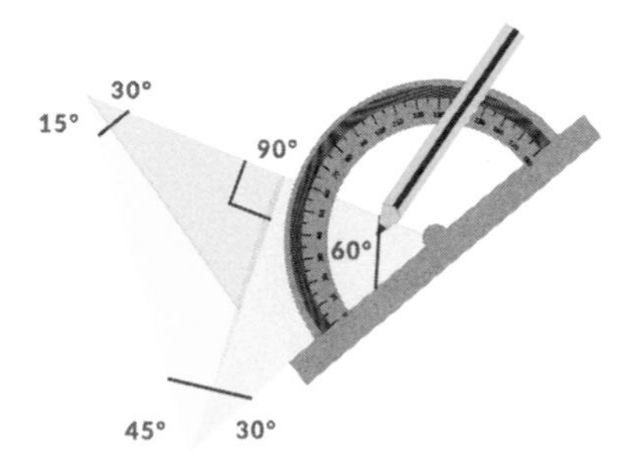

Now you can measure angles by checking which is the closest match to the ones on your protractor. Have fun!

GAMES AND PUZZLES

Math is all fun and games until someone gets stumped...

1 Triangular Reasoning

As it is now, if you add up the four numbers on each side of the large triangle below, you'll get three different sums: 18, 16, and 22. But what you want is for each side to equal the same amount. You can accomplish that by swapping two pairs of numbers. What are the swaps, and what number do all the sides add up to?

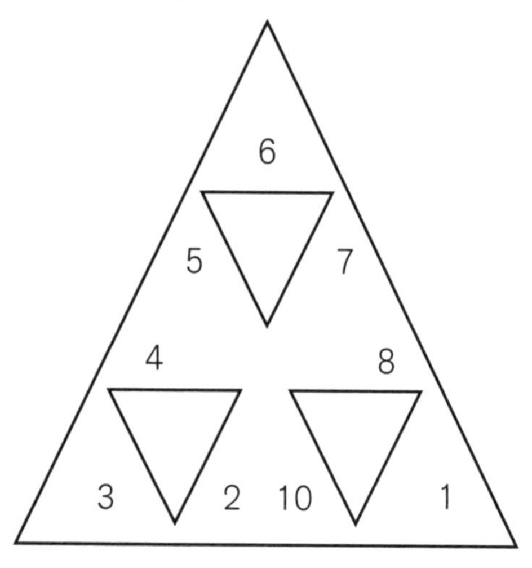

2 Mystery Number

What number are these clues describing?

1. Add its digits together and you get a prime number (a number that's divisible only by 1 and itself).

2. Subtract its first digit from the second and you get 5.

3. The number is less than the number of days in February.

3 Subtracting Up

Can you figure out a way to take away 1 from 14 and end up with 15?

4 And the Number Is...?

If a full bottle of water weighs 18 ounces and the water weighs twice as much as the bottle, how much does the bottle weigh?

5 It's Symbolic

Put a mathematical symbol between the numbers 4 and 6 to get a number that's less than 6 but more than 4.

6 Quick Trick

Move one number to make this equation correct:

$$101 - 102 = 1$$

7 A Berry Good Puzzle

Hoppy the hobo was drifting through the countryside when he spied a strawberry farm. He was hungry, so he swiped as many berries as he could carry in his pockets. Suddenly, the farmer and his two sons appeared. Hoppy was caught red-handed and, with no way to explain himself, did the first thing that came to mind: he gave the farmer half of the loot.

But that wasn't enough! After getting stern looks from the three men, he gave one son half of the remaining berries, and the other son half the berries left after that. To be on the safe side, he gave all three of them one more berry apiece. There were only two berries left, and the farmer, in his generosity, let Hoppy keep those.

How many berries did the hobo give back?

8 Oink, Oink, Ouch!

You've just smashed open your piggy bank to discover 55 coins totaling $10.00. If there are more nickels than pennies, more dimes than nickels, and more quarters than dimes, how many of each coin do you have?

9 Mix It Up

Use all of these numbers (1, 2, 3, 4, 5, 6, 9) in any combination, but only one time each, to make the following equation work from left to right (ignore the proper order of operations you learned on page 101):

? – ? x ? = ?

10 Preferences

John likes 25 but not 26, 196 but not 195, and 100 but not 99. Does he prefer 63 or 64?

11. Logic by the Numbers

To solve this one, you'll need only a glancing familiarity with math—this one can be solved mostly by logic. There's a two-digit number that, when read from right to left, is $4\frac{1}{2}$ times as large as from left to right. Can you figure out what it is? Here are some hints if you get stuck:

· The number is greater than 9 because it has two digits.

· The number is less than 23 because 23 x $4\frac{1}{2}$ is greater than 100 (a three-digit number).

· The number is an even number because multiplying an odd number by $4\frac{1}{2}$ will produce a fraction or decimal.

· Half the number times 9 is its reverse, so its reverse is divisible by 9.

12 Brotherly Love

The Hurley gang consists of three brothers—Earl, Merle, and Burly—who were suspects in three recent robberies. According to the results of the police statements and eyewitness accounts described below, which brothers committed which crimes?

1. Two of the brothers carried out the Hiawatha Casino caper.

2. Two of them broke into the Adorable You Boutique.

3. Two of them raided the Blink's armored truck.

4. The one who wasn't in on the Blink's job wasn't involved in the Adorable You Boutique break-in.

5. Burly wasn't at the Adorable You Boutique break-in or the Hiawatha Casino caper.

13 Missing Numbers

In the problem below, replace each question mark with either 1, 2, 3, or 4 to make the numbers add up correctly. Each number is used only once.

```
    ?  6  ?  7
 +  5  8  4  ?
 ------------
    7  ?  7  9
```

14 Circular Reasoning

Insert the missing number.

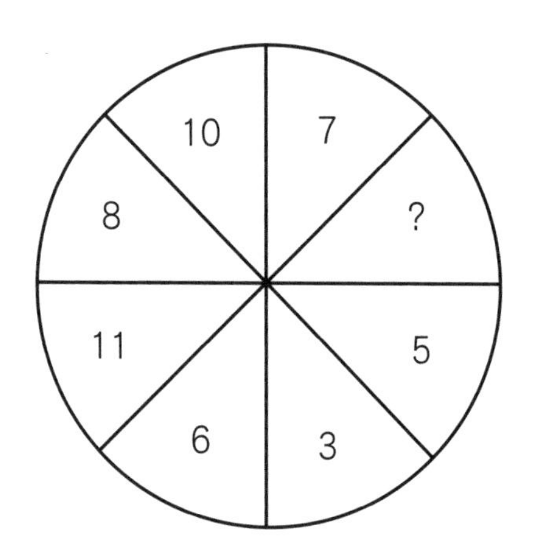

15 Above or Below?

The numbers below have been placed according to a particular rule. Can you figure out where to put the 9 and 10?

1	4	7	
2	3	5	6

16 Odd Man Out

Study the following lists of numbers and identify the one that doesn't belong in each series.

1. 64818, 93618, 54716, 63120, 82313

2. 1623, 9364, 6023, 8421, 6349, 3612, 3062

3. 13, 11, 1255, 273, 5128

4. 4369, 8609, 8162, 6122, 5306

17 Happy Couples

Old man Grouchbottom left $100,000 in his will to six beneficiaries: his three sons and their wives. The three wives in total receive $39,600, of which Nell gets $1,000 more than Ursula, and Helena gets $1,000 more than Nell. Of the sons, Mark gets twice as much as his wife, Lonnie gets the same as his wife, and Edwin gets 50 percent more than his wife.

Who is married to whom?

18 Threes Please

Use addition, subtraction, multiplication, or division to make these equations correct. All you have to do is insert the right symbols and work out the equation from left to right (in other words, ignore the proper order of operations).

1. 3 3 3 3 3 = 24

2. 3 3 3 3 3 = 0

3. 3 3 3 3 3 = 6

19 Trixie's Kids

The last census taker who'd knocked on Trixie Lott's front door had gone away shaking his head and retired from the business soon after. This time around, though, the census taker was also a lover of puzzles.

First, he asked Trixie how many children she had and got the reply "Three." But when he asked for their ages, Trixie refused to say. "I'll give you a hint," she said. "If you multiply their three ages, you get 36. And, by the way, their ages are all whole numbers." The census taker thought for a moment, then asked for another hint. So Trixie said, "The sum of their ages is the number on the house next door." After checking the number, the census taker asked for one final hint. When Trixie answered, "Tomorrow is my oldest child's birthday," the census taker jotted down a few numbers and came up with the right answer.

What are the ages of Trixie's three children?

Math-Minded + Famous

- **Michael Jordan majored in math for two years in college.**
- **Actor / wrestler Mr. T also majored in math.**
- **Brian May was a math teacher before becoming lead guitarist for Queen.**

SOLUTIONS

1 Triangular Reasoning

First swap the 5 and 7 in the second row. Then swap the 7 with the 3 in the bottom left corner. Each side will now add up to 20.

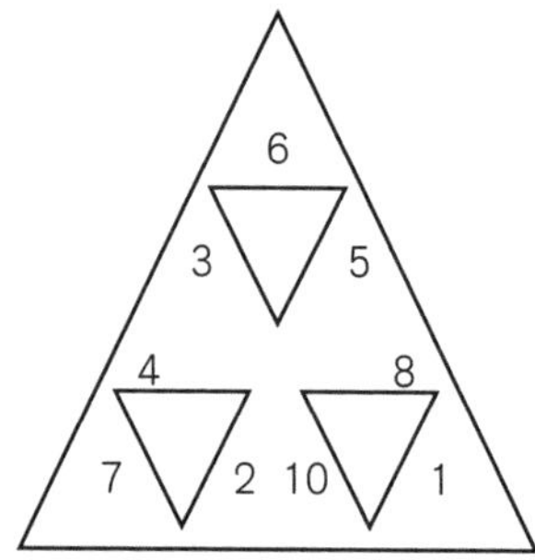

2 Mystery Number

16.

3 Subtracting Up

Use Roman numerals. If you take away I from XIV, you're left with XV.

4 And the Number Is...?

The bottle weighs 6 ounces; the water weighs 12 ounces.

5 It's Symbolic

Use a decimal point: 4.6

6 Quick Trick

$101 - 10^2 = 1$

7 A Berry Good Puzzle

38. (He took 40 total.)

8 Oink Oink Ouch!

You have 35 quarters ($8.75), 9 dimes (90 cents), 6 nickels (30 cents), and 5 pennies (5 cents).

9 Mix It Up

52 – 3 x 4 = 196

10 Preferences

He prefers 64. John likes numbers that are perfect squares.

11 Logic by the Numbers

18 is the only even multiple of 9 between 10 and 22. Check it out: $18 \times 4\frac{1}{2} = 81$

12 Brotherly Love

The same two brothers were involved in the Blink's armored truck job and the Adorable You Boutique break-in. Burly was not at the Adorable You break-in, so he was not at the Blink's job. And since he was not at the Hiawatha Casino caper, he took part in none of the robberies. Earl and Merle committed all three.

13 Missing Numbers

1637 + 5842 = 7479

14 Circular Reasoning

2. When added together, the numbers opposite each other equal 13.

15 Above or Below?

The 9 and 10 go below the line. Numbers with curved lines go below the line; those with straight lines go above.

16 Odd Man Out

1. 63120. In the other four, the first three numbers added together make the last two.

2. 8421. The other six are pairs in which the third and second digits are followed by the fourth and first.

3. 13. In the other four, cube the last digit to get the first ones.

4. 8609. In the other four, the first and last numbers multiplied together give you the second and third.

17 Happy Couples

Nell is married to Edwin, Ursula is married to Lonnie, Helena is married to Mark.

18 Threes Please

1. $3 + 3 + 3 \times 3 - 3 = 24$
2. $3 - 3 + 3 - 3 \times 3 = 0$
3. $3 \times 3 - 3 \times 3 \div 3 = 6$

19 Trixie's Kids

Knowing that their ages multiplied together equaled 36, the census taker knew their ages were one of these eight possibilities:

$$1 \times 1 \times 36 = 36$$

$$1 \times 2 \times 18 = 36$$

$$1 \times 3 \times 12 = 36$$

$$1 \times 4 \times 9 = 36$$

$$1 \times 6 \times 6 = 36$$

$$2 \times 2 \times 9 = 36$$

$$2 \times 3 \times 6 = 36$$

$$3 \times 3 \times 4 = 36$$

Each set of numbers added up equal the following totals:

$$1 + 1 + 36 = 38$$

$$1 + 2 + 18 = 21$$

$$1 + 3 + 12 = 16$$

$$1 + 4 + 9 = 14$$

$$1 + 6 + 6 = 13$$

$$2 + 2 + 9 = 13$$

$$2 + 3 + 6 = 11$$

$$3 + 3 + 4 = 10$$

Because 13 is the only sum with two possibilities, the census taker had to ask for another hint. (If the number on the house next door had been any of the other choices, the census taker wouldn't have needed a third hint.) So, the house number must have been 13. When Trixie said that tomorrow was her oldest child's birthday, the census taker ruled out the 1, 6, 6 combination because it has no largest number, and knew for certain that Trixie's kids were 2, 2, and 9 years old.

CHAPTER WRAP-UP

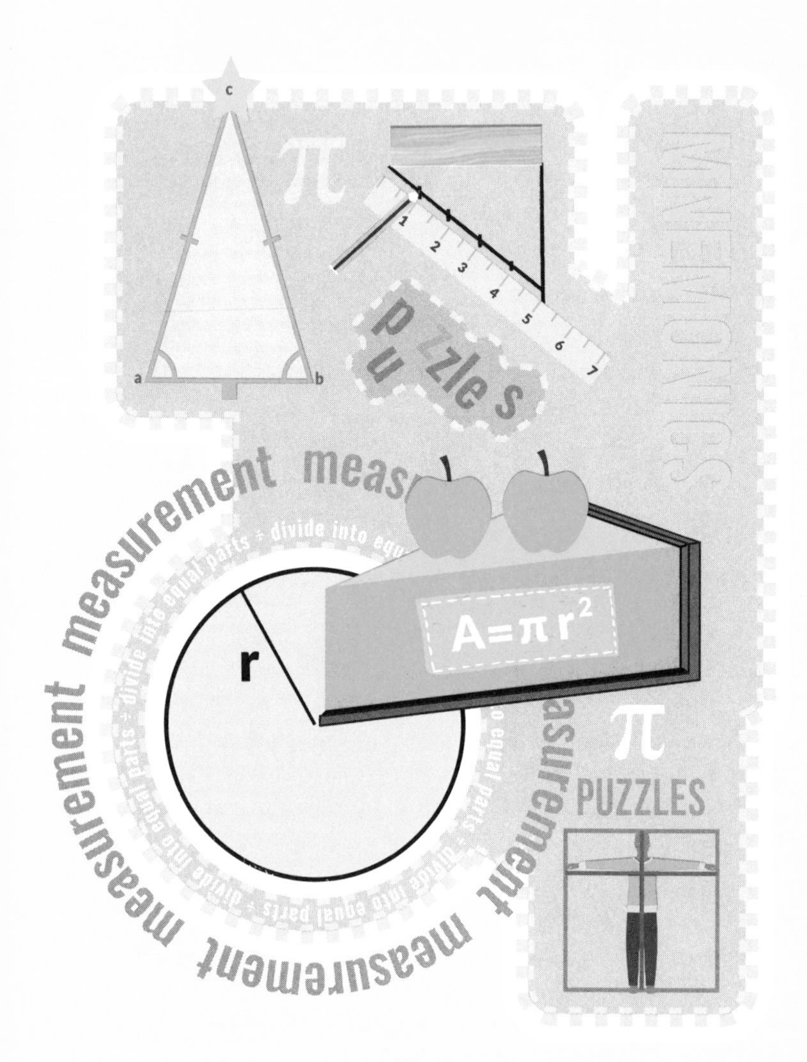

GLOSSARY

Algebra: The branch of math in which letters and symbols are used to represent numbers

Circumference: The distance around a circle

Denominator: The bottom number of a fraction

Diameter (of a circle): The distance from a point on a circle across the center to the opposite point

Dividend: In a division problem, it's the number that is being divided. Example: In 128 / 4, the dividend is 128.

Divisor: In a division problem, it's the number by which the other number is being divided. Example: In 128 / 4, the divisor is 4.

Mixed number: A number that has both a whole number and a fraction. Example: $2\frac{1}{3}$

Numerator: The top number of a fraction

Pi, or π: The ratio of a circle's circumference to its diameter; it is always the same number, a neverending number that starts with 3.14.

Principal: an initial investment

Radius: Half of a circle's diameter; the distance from the center outward to any point on the circle

Right angle: A 90-degree angle

Right triangle:
A triangle that has a 90-degree angle

Square (a number):
A number squared is that number multiplied by itself. Example: 5^2 is 5 x 5, or 25.

Square root:
A number that is squared (multiplied by itself) is called the square root of the larger number you get. Using the previous example, 5 is a square root of 25.

Whole number:
A number with no fractions or decimals. Example: 287

> **"Numbers written on restaurant bills within the confines of restaurants do not follow the same mathematical laws as numbers written on any other pieces of paper in any other parts of the Universe."**
>
> **–Douglas Adams**